STATES OF MATTER

STATES OF MATTER

By

Dr. A. Goel

DISCOVERY PUBLISHING HOUSE
NEW DELHI-110002

First Published-2006

ISBN 81-8356-131-4

Published by

DISCOVERY PUBLISHING HOUSE

4831/24, Ansari Road, Prahlad Street,
Darya Ganj, New Delhi-110002 (India)
Phone: 23279245 • Fax: 91-11-23253475
E-mail:dphtemp@indiatimes.com

Printed at:

Amit Enterprises

Preface

This book has been written for the students of under-graduate and post-graduate level of the various universities in India. A special feature of the book is that the text has been illustrated with a large number of line diagrams and the data presented in the form of numerous tables for reference and comparison. In the preparation of text standard works and review by renowned author have been freely consulted and the reference given chapter wise. At the end of the book will be found useful by those who wish to make a more detailed study of the topics discussed.

We are extremely grateful to our respected Managing Director Shri Tilak Wasan, Discovery Publishing House, for valuable and active cooperation. We humble request our students and chemistry teachers to send us their constructive criticism and suggestions which we shall be using in the publication of the next edition.

Author

Contents

Boyle's Temperature, Liquefaction of Gases, Critical Constants, Law of Corresponding States, Other Equations of State, Non-Uniform Gases, Thermal Conductivity, Transport Properties, Glassy State, Inorganic Glasses, Devitrification, Viscosity Temperature Characteristics, Second Order Transition, Transformation Temperature, Relaxation, Mechanism of Glass Transition, Methods of Determining T_g, Structure, Organic Glasses, Glass Transition Temperature and Molecular Mass, Glass Transition Temperature and Chemical Constitution of Polymers, Forced Elasticity, Stress Strain Dependence, Brittle Temperature, Effect of Intermolecular Attraction Energy, Effect of Packing in Macromolecules, Effect of molecular Mass, Glass Transition by Chromatography, Plastic State, Plastics, Classification According to Mechanical Properties, Resins, Crystallite, Micelles, Monomer, Polymerization, Oligomers, Polymers (Macromolecules), Structural Unit or Radical, Functionality, High Polymers, Types of High Polymer, Ionic Polymerization, Block Complymer, Polymer Crystals, Crystallization of Polymers, Fibril, Film or Sheets, Spherulites, Globular Crystals, Shape of the Macromolecule, Craft Copolymer, Random Walk Model, Elastomers.

1

LIQUID STATE

INTRODUCTION

The liquid state lies between the gaseous and the solid state in the sense that there is neither the ordered arrangement of constituents nor the complete disorder as in gases. It is generally observed that some of the properties of liquids closely resemble those of gases while some of the properties approach those of the solids. In the light of the kinetic molecular theory, liquids may be regarded as a continuation of gases into the region of small volumes and high intermolecular attraction. The cohesive forces in a liquid have been stronger than those in gases even at high pressures, and are sufficiently high to keep the molecules confined to a definite volume. The positions of the molecules in liquids arc not rigidly fixed. These forces are not strong enough to entirely eliminate the movements of the molecules in the liquid. Hence, the molecules in the liquid state have much shorter mean free path as compared to gas molecules.

Properties of liquids can be explained on the basis of their following characteristics :

(a) Appreciable forces of attractions exist between the molecules of a liquid. These are about 10^6 times as strong as in gases. The forces amongst the constituents of a liquid disallow them from separating spontaneously from each other, but they are not strong enough to hold the molecules in fixed positions. It is possible to explain surface tension, viscosity, fairly high heat of vaporization and vapour pressure of liquids in terms of these attractive forces.

(b) The molecules in a liquid are in a state of random motion although the extent of randomness is appreciably much smaller in comparison to gases. The molecules are relatively close together. Most of the space in the liquid gets occupied by its molecules and only a small fraction of the space is available to them for their free movement. This explains the higher density, incompressibility and slow diffusion of liquids in comparison to gases.

(c) The average kinetic energy of the molecules in a liquid is proportional to the absolute temperature. Increase in temperature would increase the proportion of the energized molecules, lowers the attractive forces between the molecules, and consequently increases the vapour pressure of the liquid.

From me above arguments, it is evident that most of the characteristic properties of the liquid arise due to the nature and magnitude of the intermolecular forces between the molecules. The important properties dealt with in this chapter are :

(i) Vapour pressure,

(ii) Surface tension, and

(iii) Viscosity.

LIQUID-VAPOUR EQUILIBRIUM-VAPOUR PRESSURE

Molecules of a liquid, like those of a gas, exhibit a distribution of energies. The molecules that are having sufficient energy to overcome intermolecular attractions are able to escape from the bulk into the gas phase (or vapour phase) provided they are close to the surface. This happens during evaporation. But the average kinetic energy of molecules in vapour state is more than that in liquid state. Therefore the temperature of the liquid falls on evaporation. The rate of evaporation of a liquid depends upon :

(a) The temperature of the liquid,

(b) Attractive forces in the liquid,

(c) Surface area, and

(d) Pressure above the liquid.

If we examine the evaporation of a liquid in an enclosed space (at constant temperature), it is found that some of the molecules of the liquid will spontaneously pass from the surface into the space above it (Fig. 1.1).

But the space is closed. Therefore, the molecules in the vapour state are unable to escape from the container. The molecules in the vapour phase collide with each other and with the sides of the container. Some of the molecules on collision pass on their energy to other molecules and come back to the liquid phase. This phenomenon is termed as condensation. The rate of condensation of molecules in the vapour phase is proportional to the concentration of molecules in the vapour phase. A stage is ultimately reached when the rate of condensation becomes equal to the rate of evaporation. This situation corresponds to that of a dynamic equilibrium between the liquid and its vapour, *e.g.*,

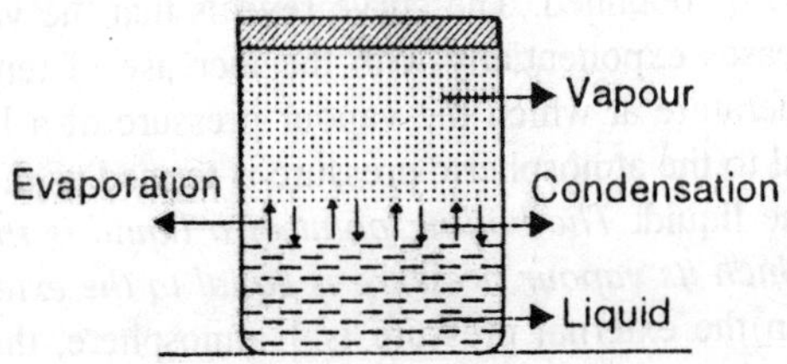

Fig. 1.1 : Liquid-vapour equilibrium.

$$H_2O(l) \underset{\text{condensation}}{\overset{\text{evaporation}}{\rightleftharpoons}} H_2O(g)$$

Thus a state of equilibrium may be defined as one in which two opposing processes occur simultaneously at the same rate. At this point there is no further change in the number of molecules in either the liquid or in the vapour state.

Molecules of a liquid in the vapour state exert some pressure. At equilibrium, this pressure is characteristic of the liquid at a given temperature and is called the vapour pressure of the liquid. The vapour pressure of a liquid is therefore defined as the pressure exerted by the vapours that are in equilibrium with the liquid at a given temperature. The vapour pressure has been found to depend on

(i) The nature of the intermolecular forces in the liquid, and

(ii) The temperature of the liquid.

(i) *Nature of the Liquid :* If in a liquid the attractive forces between the molecules are strong then the tendency of the molecules to escape from the surface of the liquid would be low. Hence, such a liquid will have low vapour pressure. However, liquids with low intermolecular forces would be having a high escaping tendency and consequently high equilibrium vapour pressure.

(ii) *Effect of Temperature* : As the temperature of a liquid gets increased the average kinetic energy of the molecules increases. The increased kinetic energy partly overcomes the attractive forces between the molecules and thus raises the escaping tendency of the molecules. This causes an increase in the equilibrium vapour pressure. The values of vapour pressure when plotted against temperature, a curve of the type shown in Fig. 2 is obtained. The curve reveals that the vapour pressure increases exponentially with the increase of temperature. The temperature at which the vapour pressure of a liquid becomes equal to the atmospheric pressure is termed as the boiling point of the liquid. *The boiling point of a liquid is the temperature at which its vapour pressure is equal to the external pressure.* When the external pressure is 1 atmosphere, the term normal boiling point is used.

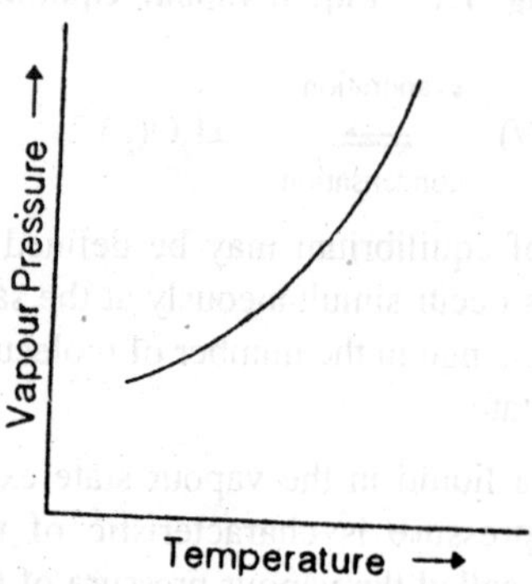

Fig. 1.2 : Variation of vapour pressure of a liquid with temperature.

METHODS OF MEASURING VAPOUR PRESSURE OF LIQUIDS

The various methods used for measuring vapour pressures of liquids may be classified into two categories, viz., *static and dynamic* methods. In the *static methods* the liquid is allowed to establish its vapour pressure without being disturbed while in the *dynamic methods* the liquid is either boiled or has a stream of inert gas passing through it.

The Static Methods

1. Barometric Method

It involves the use of two barometers as shown in Fig. 1.3 one of which is used for comparison. The given liquid is introduced into one of the tubes by means of a teat pipette, until the space above mercury is saturated with vapour, as shown by a small amount of liquid remaining on the surface of mercury. The difference in levels of mercury in the two tubes gives the vapour pressure of the liquid at the given temperature T. Measurements can be made at a series of temperatures by enclosing the two tubes in a jacket maintained at different temperatures by circulating a suitable liquid.

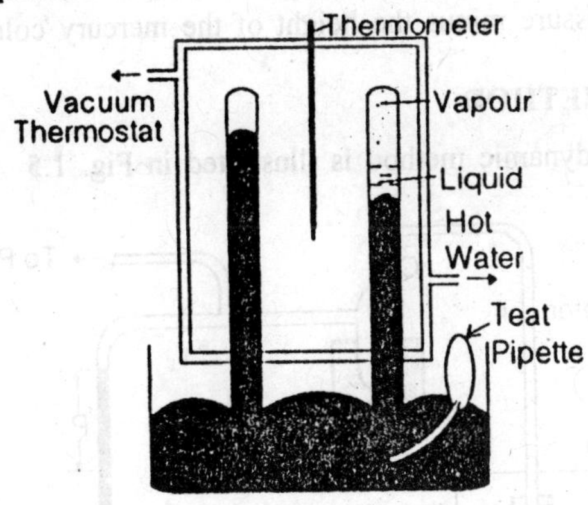

Fig 1.3

2. Isoteniscopic Method

A useful form of the Static method is the isoteniscopic method devised by A. Smith and *A.W.C. Menzies* in 1910.

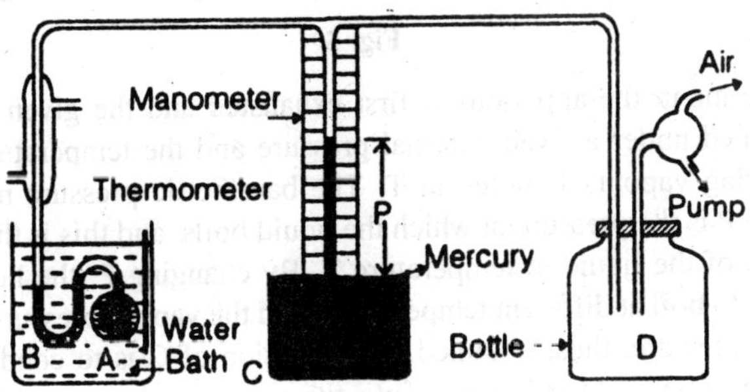

Fig 4

A bulb A, of approximately 2 cm diameter, is about half-filled with the given liquid and the U-shaped portion of the tube B, is filled to a depth of 2 or 3 cm with the same liquid. The isoteniscope A-B is then surrounded by a water bath maintained at a constant temperature T. The U-tube is connected to a manometer C and a large bottle D to smooth out pressure fluctuations in the system. The bottle can be connected alternately to a suction pump or to air. The apparatus is evacuated until the liquid boils vigorously and the air in A is removed. Air is now slowly admitted into the system until the liquid levels in the U-tube B are exactly equal. At this stage the pressure on either side of the U-tube becomes equal. The pressure above the liquid in A, *i.e.*, the vapour pressure at temperature T, is then equal to the pressure in D. This is equal to the barometric pressure minus the height of the mercury column in C.

DYNAMIC METHOD

A simple dynamic method is illustrated in Fig. 1.5

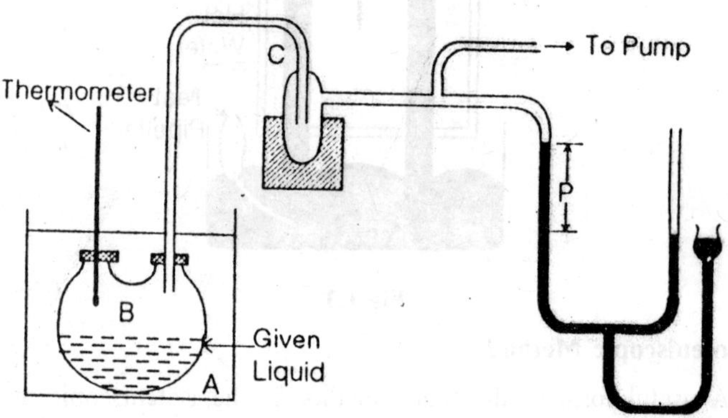

Fig. 5

The air in the apparatus is first exhausted and the given liquid in B is boiled under a fixed external pressure and the temperature of the condensing vapours is noted at T. The barometric pressure minus the pressure P is the pressure at which the liquid boils, and this is the vapour pressure of the liquid at temperature T. By changing P, the liquid may be made to boil at different temperatures and the vapour pressure at these temperatures are, thus, obtained. The function of C is to condense any escaping vapours from passing into the manometer.

EQUATION OF STATE FOR LIQUIDS

In gases the distance between the molecules is large as compared to that in solids or liquids. All gases behave nearly alike at very low pressures (strictly speaking $P \rightarrow 0$) so that their behaviour can be adequately explained by an equation of state, $PV = nRT$. Solids and liquids are considered as *condensed phases*. Because of relatively small distances between the molecules, the effects of intermolecular forces are quite predominant in solids and liquids. The magnitude of these forces tend to vary with me nature of the liquid or the solid. Thus, a general equation of state for condensed phases would not be possible.

An equation of state for solids and liquids can be deduced in terms of coefficients of thermal expansion (a) and compressibility (k). These are defined as

$$\alpha = \frac{1}{V}\left(\frac{\partial V}{\partial T}\right)_P \qquad ...(1)$$

$$k = -\frac{1}{V}\left(\frac{\partial V}{\partial T}\right)_T \qquad ...(2)$$

where α refers to the isobaric relative increase $\left(\frac{\partial V}{V}\right)$ in volume per degree rise in temperature while k isothermal relative decrease $\left(-\frac{dV}{V}\right)$ per unit rise in pressure, a and k depend on the nature of the condensed phases and are different for different substances.

For small temperature intervals equation (1) can be written as follows :

$$\frac{dV}{V} = \alpha \, dT$$

Integrating the above equation within limits, we get

$$\int_{V_0}^{V_t} \frac{dV}{V} = \int_0^T dT$$

$$\text{In}\,\frac{V_t}{V_0} = \alpha t$$

or $\qquad V_t = V_0 \, e^{\alpha t}$

where V_t and V_0 refer to the volumes at t° and 0°C respectively. Expanding $e^{\alpha t}$ in the power series, we get

$$e^{\alpha t} = 1 + \alpha t\left(\frac{(\alpha t)^2}{2!}\right) + ...$$

As α is very small, square and other higher order terms may be neglected. Therefore, we can write

$$V_t = V_0(1 + \alpha t) \qquad ...(3)$$

Equation (3) relates the volume of a substance with temperature, α is constant over a limited temperature range and is approximately constant for gases.

In case of liquids or solids, α is having its own characteristic value, α is having a positive value for gases and solids while for most of the liquids it is positive with some exception, *e.g.*, water between 0 – 4°C has negative value of α.

Similarly, integrating equation (2) we obtain

$$\int_{V_0^0}^{V0} \frac{dV}{V} = -\int_1^P kdP$$

$$\ln \frac{V_0}{V_0^0} = -k(P-1)$$

or $$V_0 = V_0^0 e^{-k(P-1)}$$

where V_0^0 refers to the volume at 1 atmosphere and 0°C.

On expanding $e^{-k(P-1)}$ in the power series and neglecting square and other higher terms, we obtain

$$V^0 = V_0^0 [1 - k(P - 1)]$$

Like α, k is always positive and is the same for all gases but each liquid or solid has its own characteristic value of k. k is constant over a fairly large range of pressure. Equation (4) predicts that the volume of a condensed phase gets decreased linearly with increase of pressure.

This behaviour is contrary to the behaviour of gases in which volume is inversely proportional to pressure. Combining equations (3) and (4) and eliminating V_0, we obtain

$$V_t = V_0^0 (1 + \alpha t) [1 - k(P - 1)] \qquad ...(5)$$

The constants a and k in equation (5) depend upon the nature of the condensed phase. Table 1.1 gives the values of α and k for some common liquids and solids.

Table 1 : Coefficient of Thermal Expansion and Compressibility at 293 K.

Solid	*Coefficients*		*Liquid*	*Coefficients*	
	$\alpha\times10^{4}\ k^{-1}$	$k\times10^{6}\ atm^{-1}$		$a\times10^{-4}\ K^{-1}$	$k\times10^{6}\ a^{-1}$
Silver	0.582	1.00	C_6H_6	12.4	94.0
Copper	0.491	0.78	CH_3OH	12.0	120.0
Graphite	0.242	3.00	C_2H_5OH	11.2	110.1
Platinum	0.263	0.38	CCl_4	12.4	103.2
NaCI	1.212	4.20	H_2O	2.0	45.3

STRUCTURE OF LIQUIDS

In a perfect crystal a complete ordered arrangement of the atoms, ions or molecules constituting the crystal exists. The intermolecular forces are strong enough to hold them together. The constituents vibrate about their mean positions but cannot execute Translational motion. Hence both the short range as well as the long range orders exist in crystals. The molecules in a gas have complete random motion and the intermolecular forces between the molecules are small and are effective only at short distances. There is no possibility of any kind of ordered structure in gases. In liquids, on the other hand, the situation somewhat lies in between the two. The cohesive forces in liquids are stronger than those in gases. These forces, however, are not strong enough to disallow considerably the translational motion of the individual molecules. In terms of the arrangement of the constituents a liquid is having *short range order* but lacks long range order.

When a crystal melts there is, in general, an increase in volume of about 10% or about 3% in intermolecular spacings. The molecules in the liquid state still remain in the vicinity of the other molecules surrounding them. The intermolecular forces tend to decrease as the distance between the molecules increases. In other words, the short range forces exist while the long range forces are negligible. The thermal motion tends to introduce a disorder in the structure but the motion is not as random as in a gas. Some ordered arrangement still persists. The ordered arrangement in the crystal is not completely destroyed on melting. As the temperature gets further increased, the thermal motions of the molecules increase the kinetic energy, decrease the order until at the boiling point it completely vanishes.

The sharpness of the melting point could be explained by a two dimensional model for solids, liquids and gases as shown in Fig. 6. Introduction of a small region of disorder into a crystal brings about the disturbance of the long range order and destroys the crystalline arrangement. This explains the abrupt variation in the properties between solids and liquids. J.D. Bemal postulated that an atom is surrounded by five atoms (Fig. 1.6) instead of normal six atoms as is the requirement for a closest packed structure for solids. The remaining atoms (circles in the figure) are drawn in the utmost possible ordered arrangement. One point of abnormal coordination number is sufficient to bring about a long range disorder. That is, when the normal motion causes a disorder in one region, it spreads in all directions destroying the entire regular structure. Liquids, therefore, may be regarded to have structure like that of a crystal with a difference that the ordered arrangement extends over a short region instead of over the whole mass. This is termed as the short range order or long range disorder. It is to be noted that the short range order in a liquid structure is continuously changing because of the thermal motions.

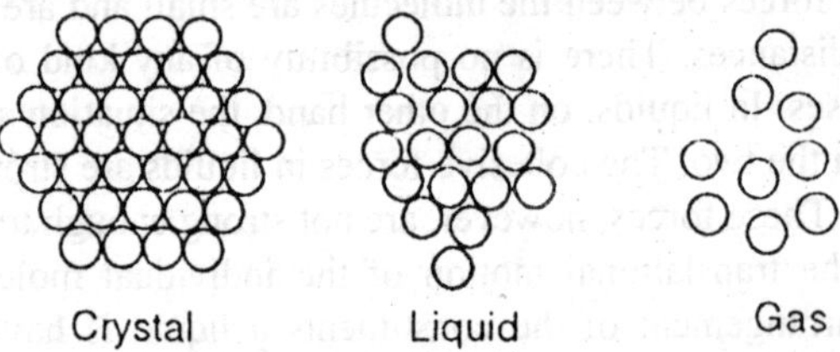

Fig. 1.6 : Two dimensional models of states of matter.

X-ray Diffraction of Liquids : It becomes possible to calculate the distribution of atoms or molecules in a liquid by the analysis of the intensity of the scattered X-rays. The diffraction pattern of a liquid resembles a powder photograph with a difference that in place of sharp lines a few broad bands are observed. This reveals the existence of a regular order. From the intensities of these bands, it is possible to construct the radial distribution function which is interpreted in terms of the average number of atoms around a central atom at a certain distance. As the temperature of the liquid gets raised, the bands become less pronounced and the pattern resembles those of a gas revealing a more disordered arrangement at higher temperatures.

Vacancy Model for a Liquid : This model for the structure of liquids was given by Eyring (1933) who assumed the liquids more or less similar to a gas. In the liquid, most of the space gets occupied by the molecules and only a small fraction (about 3% of the total volume at ordinary temperature and pressure) of the total volume is free or void. This void space in which the molecules can move is known as the *free volume.* This is shown in Fig. 1.7. The vapours have only a few molecules moving randomly and thus has large free volume. With a rise in temperature, the concentration of the molecules in the vapour phase increases bringing about an increase in the vacancies of the liquid. The density of the vapour increases while that of the liquid decreases until they become equal at the critical temperature. At the critical temperature the liquid and vapour become indistinguishable. Eyring further regarded that the free volume gets distributed randomly and the vacancies are of approximately molecular size. Molecules adjacent to a vacancy would be expected to posses gas-like properties and show maximum disorder. But a molecule away from the vacancy would possess solid-like properties and possessing a more ordered arrangement. Based on this model, they were able to show a fairly good agreement between the observed and predicted values of the properties.

Radial Distribution Function : The particles of a liquid are held together by intermolecular forces, but their kinetic energies are comparable to their potential energies. As a result, the whole structure is very mobile. The mathematical description of the average locations of the particles is expressed in terms of the radial distribution function g, this function is defined so gr^2 dr is the probability that a molecule will be found in the range dr at a distance r from another.

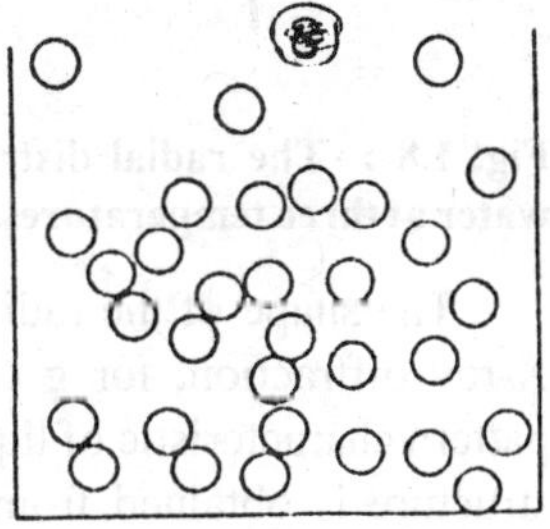

Fig. 7 : Eyring's vacancy model of liquids.

In a perfect crystal, g is a periodic array of sharp spikes, representing the certainty (in the absence of defects and thermal motion) that particles lie at definite locations. This regularity continues out to the edges of the crystal, so we say that crystals have long-range order. When the crystal melts, the long-range order is lost and, wherever we look at long distances from a given particle, there is equal probability of finding a second particle. Close to the first particle, though, the nearest neighbours might still adopt approximately their original relative positions and, even if they are displaced by new-comers, the new particles might adopt their vacated positions. It is still possible to detect a sphere of nearest neighbours at a distance r_1 and perhaps beyond them a sphere of next-nearest neighbours at r_2. The existence of this short-range order means that the radial distribution function can be expected to oscillate at short distances, with a peak at r_1 a smaller peak at r_2 and perhaps some more structure beyond that.

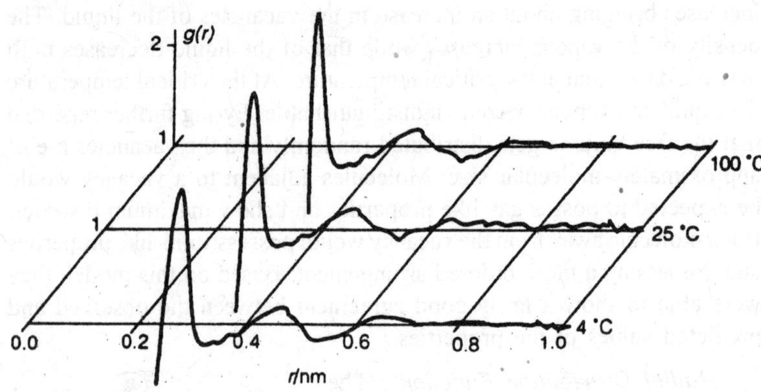

Fig. 1.8 : The radial distribution function of the oxygen atoms in liquid water at three temperatures. Note the expansion as the temperature is raised.

The shape of the radial distribution function can be determined by X-ray diffraction, for g can be extracted from the diffuse diffraction pattern characteristic of liquid samples in much the same way as a crystal structure is obtained from X-ray diffraction of crystals. The shells of local structure shown in the example in Fig. 1.8 (for water) are unmistakable.

Closer analysis shows that any given H_2O molecule is surrounded by other molecules at the comers of a tetrahedron, similar to the arrangement in ice. The form of g at 100°C shows that the intermolecular forces (in this case, largely hydrogen bonds) are strong enough to affect the local structure right up to the boiling point.

Calculation of g

Because the radial distribution function can be calculated by making assumptions about the intermolecular forces, it can be used to test theories of liquid structure. However, even a fluid of hard spheres without attractive interactions (a collection of ball-bearings in a container) gives a function that oscillates near the origin (Fig. 1.9), and one of the factors influencing, and sometimes dominating, the structure of a liquid is the geometrical problem of stacking together reasonably hard spheres. Indeed, a liquid of hard spheres shows a more pronounced oscillatory behaviour at a given temperature than any other type of liquid. The attractive part of the potential modifies this basic structure but sometimes only quite weakly. One of the reasons behind the difficulty of describing liquids theoretically is the importance of both the attractive and repulsive (hard core) components of the potential.

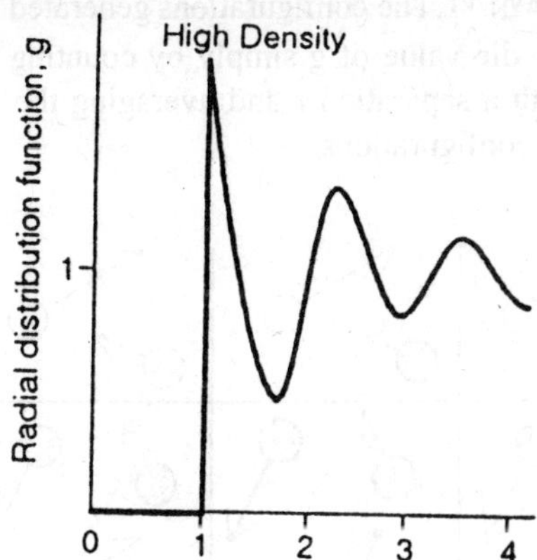

Fig. 1.9 : The radial distribution function for a simulation of a liquid using impenetrable hard spheres (ball bearings).

There are several ways of building the intermolecular potential into the calculation of g. Numerical methods take a box of 10^3 particles (the number increases as computers grow more powerful), and the rest of the liquid is simulated by surrounding the box with replications of the original box (Fig. 1.10). Then, whenever a particle leaves the box through one of its faces, its image arrives through the opposite face so that the total number remains constant.

Monte Carlo Methods

In the Monte Carlo method, the particles in the box are moved (usually one at a time) through small but otherwise random distances, and the total potential energy is calculated using one of the intermolecular potentials. Whether or not this new configuration is accepted is then judged from the following rules :

(1) If the potential energy is not greater than before the change, then the configuration is accepted.

(2) If the potential energy is greater than before the change, then it is accepted or rejected with the probability of acceptance in proportion to the value of $e^{-\Delta V_N/kT}$, where ΔV_N is the change in total potential energy of the N particles in the box.

It can be shown that this procedure ensures that at equilibrium the probability of occurrence of any configuration is proportional to the Boltzmann factor $e^{-\Delta V_N/kT}$.The configurations generated in this way can then be analyzed for die value of g simply by counting the number of pairs of particles with a separation r and averaging the result over the whole collection of configurations.

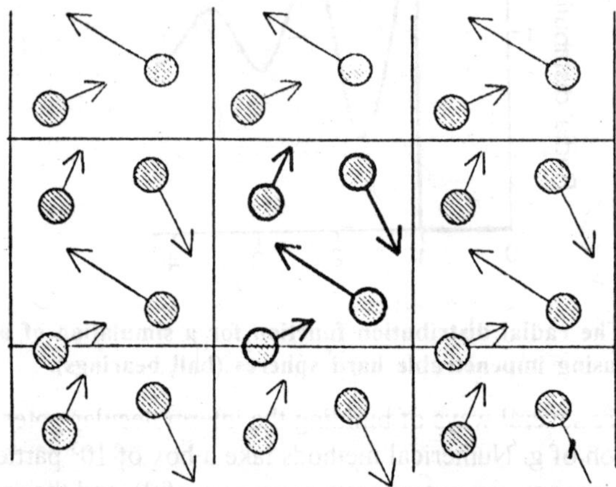

Fig. 1.10 : In a two-dimensional molecular dynamics simulation that uses periodic boundary conditions, when one particle leaves the cell its minor image enters through the opposite face.

SURFACE TENSION

A molecule in the interior of a liquid is completely surrounded by other molecules and hence, on an average, it is attracted equally in all directions. However, a molecule on the surface of a liquid is subjected to a resultant inward pull because there is a greater number of molecules per unit volume in the liquid than in the vapour. Because of this inward pull, the surface of a liquid tends to contract to the smallest possible area and behaves as if it were in a state of tension.

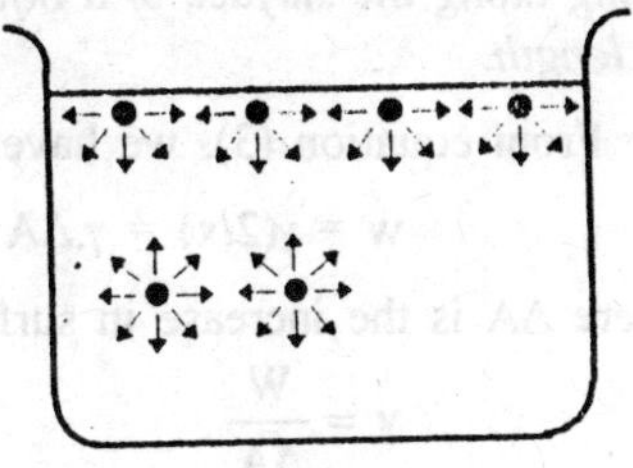

Fig. 1.11

Since the natural tendency of a liquid is to decrease its surface area, any increase in the latter can only be accomplished at the expense of work. Consider a liquid film contained within a rectangular wire frame PQRS, as shown in Fig. 1.12.

The side RS, of length l is movable. If a force 'f' is required to move RS against the force of surface tension acting in the film along RS, then the work, W, in moving the wire from RS to AB is given by

$$W = f.x \qquad ...(1)$$

This force f will have to be balanced by the force of surface tension along RS. If γ is the force acting per cm along RS due to the surface tension and since there are two surfaces to the film, the force due to surface tension would be $2\gamma l$, then

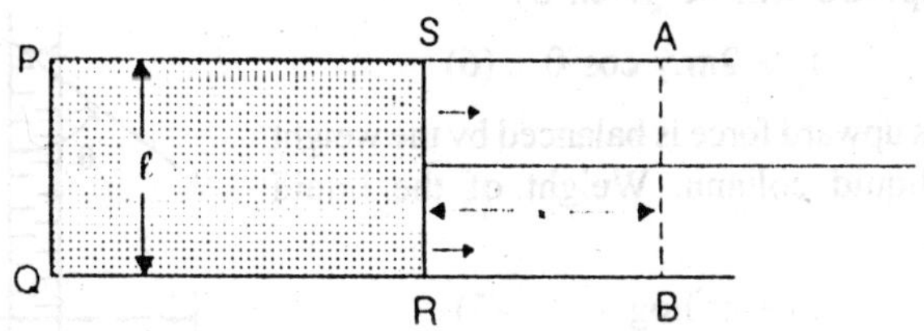

Fig. 1.12

Work done in increasing the surface area $= 2\gamma.l.x \qquad ...(2)$

or $\qquad W = 2\gamma.l..x = f.x \qquad ...(3)$

Hence, $\gamma = \frac{f}{2l}$ ...(4)

Therefore, surface tension may be defined as *the force in dyne acting along the surface of a liquid at right angles to any line one cm dn length.*

From equation (3), we have

$$w = \gamma(2lx) = \gamma.\Delta A$$

where ΔA is the increase in surface area of the film. It follows that

$$\gamma = \frac{W}{\Delta A} \quad ...(5)$$

Therefore, surface tension is the work in ergs required to generate 1 sq. cm of surface area

METHODS OF MEASURING SURFACE TENSION

The methods more frequently used in the determination of surface tension of liquids are described below.

Single Capillary Rise Method

Consider a fine capillary tube of a uniform radius 'r' immersed in a vessel containing a liquid which wets glass. The liquid will rise in the capillary till the force of surface tension f_1 acting upward is equal to the force f_2 due to the column of liquid acting downward.

Let 'γ' be the surface tension of the liquid expressed in dynes per cm of the inner circumference. This force is effective in pulling the liquid upward. Since this force is acting around the circumference, the force acting upward will be given by

$$f_1 = 2\pi r.\gamma \cos \theta \quad ...(6)$$

This upward force is balanced by the weight of the liquid column. Weight of the liquid column,

$$f_2 = \pi r^2 h\rho g \quad ...(7)$$

where 'ρ' is the density of the liquid and 'g' is the Gaseous State due to gravity.

At equilibrium both the forces become equal and the height of the liquid remains steady

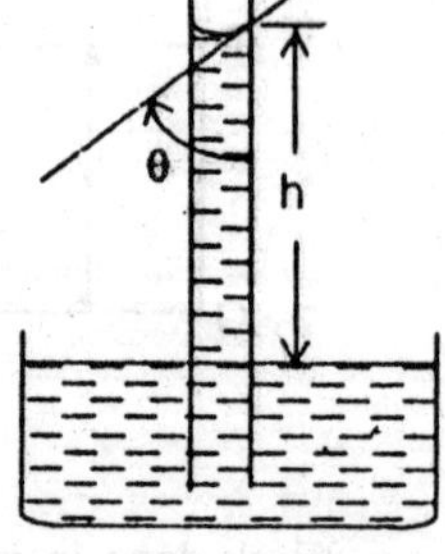

Fig. 1.13 : Rise of liquid in capillary.

Hence, $2\pi r.\gamma \cos\theta = \pi^2.h.\rho.g$

$$\therefore \quad \gamma = \frac{r h \rho g}{2 \cos\theta} \quad ...(8)$$

The angle θ is often called an angle of contact. For most liquids which wet glass, θ is nearly equal to zero.

$$\therefore \quad \cos\theta = 1$$

$$\theta \quad \gamma = \frac{r h \rho g}{2} \quad ...(9)$$

For liquids which do not wet glass, *e.g.*, mercury, the angle of contact θ has to be taken into consideration.

In order to calculate γ by means of equation (8) or (9) r has to be determined by a travelling microscope, A by means of a cathetometer and ρ by a specific gravity bottle.

Double Capillary Rise Method

This method is useful when the given liquid is available in a small quantity.

If two capillary tubes of radii r_1 and r_2 ($r_2 > r_1$), are immersed in the same liquid whose surface tension is to be determined then the liquid rises to different heights in the tubes. Let the height to which the liquid rises in the two capallaries be h_1 and h_2 cm respectively.

Assuming that θ, the angle of contact, is negligible, it follows from equation (9)

$$\text{Surface tension} = \gamma = \frac{r_1 h_1 \rho g}{2}$$

Similarly, for second capillary tube with radius r_2

$$\gamma = \frac{r_2 h_2 \rho g}{2}$$

$$\therefore \quad \frac{\gamma}{r_1} = \frac{1}{2} h_1 \rho g$$

and

$$\frac{\gamma}{r_2} = \frac{1}{2} h_2 \rho g$$

$$\therefore \quad \gamma\left[\frac{1}{r_1} - \frac{1}{r_2}\right] = \frac{1}{2}(h_1 - h_2)\,\rho g$$

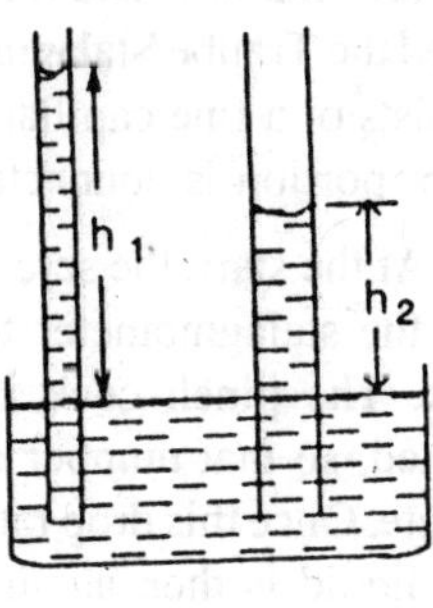

Fig. 1.14

Hence if r_1, r_2 and the difference between two heights are known, γ can be calculated from equation (10).

Drop Weight Method

The principle of this method consists in determining the weight of a single drop of a liquid which falls under its own weight through a capillary tube of uniform radius. At the instant the drop is about to fall the force f_1 due to surface tension pulling the drop upwards will be $2\pi r\gamma \cos\theta$. This force is counterbalanced by the weight of the drop f_2 acting downwards. At equilibrium both these forces become equal and hence $2\pi r\,\gamma \cos\theta = mg$, where 'm' is the mass of the drop.

Consider two different liquids having surface tensions γ_1 and γ_2 flowing through the same capillary tube. On assuming that their angles of contacts are equal, we get

$$\frac{2\pi r \gamma_1 \cos\theta}{2\pi r \gamma_2 \cos\theta} = \frac{m_1 g}{m_2 g}$$

$$\therefore \quad \frac{\gamma_1}{\gamma_2} = \frac{m_1}{m_2} \qquad \text{...(11)}$$

Hence, if we determine the masses of a single drop of each of the liquids, and knowing the surface tension of any one liquid that of the other can be calculated.

Drop Number Method

Since it is very difficult to determine the weight of a single drop, the drop weight method is not very convenient. In the drop number method, the number of drops formed from a fixed volume of a liquid (between marks A and B in Fig. 1.15) is determined by using an instrument called the Traube Stalagmometer. The lower portion of the stalagmometer consists of a fine capillary tube through which the liquids can flow. The upper portion is connected by rubber tubings to a Y-shaped glass tube.

At the start, the screw clamp is closed and the given liquid is sucked into the stalagmometer to a level above mark A by opening the pinch cock. The pinch cock is then closed and the screw clamp gradually opened, so that number of drops flowing out is between 15 and 20 per minute. Once this drop rate is achieved the screw clamp is left undisturbed. The liquid is then again sucked in by opening the pinch cock without allowing any air bubble to remain in the capillary. The number of drops formed while the liquid flows from the mark A to B is then determined.

Let this number be denoted by n_1.

If $2\pi r\gamma_1 \cos\theta$ is the force due to surface tension on a single drop, then the force on n, drops will be $2\pi r\gamma_1 n_1 \cos\theta$. This latter force will have to be balanced by the weight of n_1 drops which is M_1g where M_1 is the mass of n_1 drops.

$$\therefore \quad 2\pi r n_1\gamma_1 \cos\theta = M_1 g$$

$$\therefore \quad 2\pi r n_1\gamma_1 \cos\theta = V.\rho_1.g \quad ...(12)$$

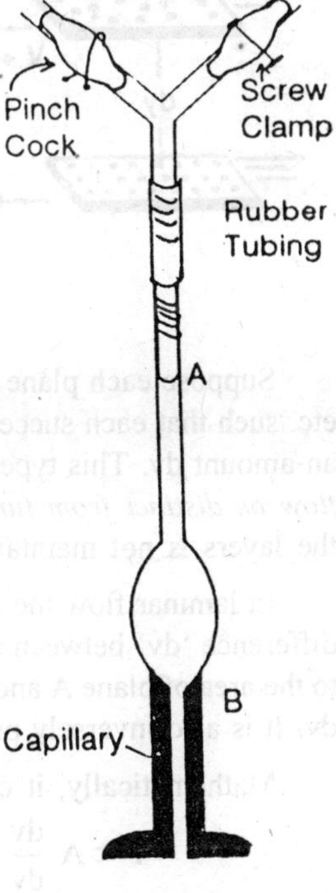

Fig. 15 : Stalagmometer.

where V = volume of the liquid between the marks A and B,

ρ_1 = density of the liquid.

The stalagmometer is then cleaned and dried and the same procedure repeated with another liquid of surface tension γ_2. Let the number of drops obtained with this liquid be denoted by n_2.

$$\therefore \quad 2\pi r\, n_2\gamma_2 \cos\theta = M_2 g$$

$$\therefore \quad 2\pi r\, n_2\gamma_2 \cos\theta = V.\rho_2.g \quad ...(13)$$

From equations (12) and (13) we have,

$$\frac{2\pi r_1\, n_1\gamma_1 \cos\theta}{2\pi r_2\, n_2\gamma_2 \cos\theta} = \frac{V.\rho_1.g}{V.\rho_2.g}$$

$$\therefore \quad \frac{n_1}{n_2} = \frac{\rho_1}{\rho_2} \times \frac{\gamma_2}{\gamma_1}$$

$$\therefore \quad \gamma_2 = \frac{n_1}{n_2} \cdot \frac{\rho_2}{\rho_1} \cdot \gamma_1 \quad ...(14)$$

Hence, if n_1 and n_2 are determined and so also the densities of the two liquids, then the surface tension of any one liquid can be calculated if that of the other liquid is known.

In addition to these methods, there are 'maximum bubble pressure' and' Torsion balance' methods which are also frequently used. The description of these methods is outside the purview of this book.

VISCOSITY

Liquids possess another property known as viscosity which implies resistance to flow. It is developed in liquids because of the shearing effect of moving one layer of liquid past another.

Consider a liquid to be divided into parallel planes or layers of molecules, a fixed distance apart. Let the area of each plane be A and the distance between planes be dy,

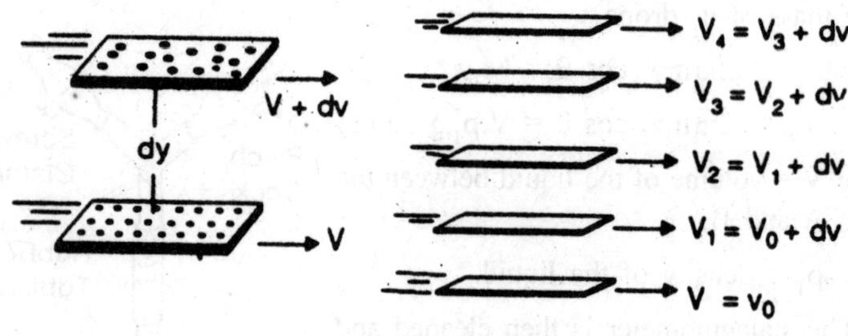

Fig. 1.16

Suppose each plane is moving to the right with velocities v_0, v_1, v_2 etc. such that each succeeding velocity is greater than the preceding by an amount dv. This type of flow is called *laminar flow* or *streamlined flow as distinct from turbulent or vortex flow* in which parallelism of the layers is not maintained.

In laminar flow the force 'f' required to maintain a steady velocity difference 'dv' between any two parallel planes is directly proportional to the area of plane A and the velocity difference between the two planes dv. It is also inversely proportional to the distance between the planes.

Mathematically, it can be expressed as

$$\therefore \quad f \propto A.\frac{dv}{dy}$$

$$\therefore \quad f = \eta.A.\frac{dv}{dy} \qquad ...(1)$$

where, η is called the coefficient of viscosity of the liquid.

$$\therefore \quad \eta = \frac{f}{A \times \frac{dv}{dy}} \qquad ...(2)$$

When A = 1 sq.cm and dv = 1cm sec^{-1},

dy = 1cm then η = force in dynes

$$\text{units } \eta = \frac{\text{dynes}}{\text{cm}^2 \times \frac{\text{cm/sec}}{\text{cm}}} = \text{dynes.sec.cm}^{-1}$$

This is the absolute unit of viscosity coefficient. The practical unit in the C.G.S. system is known as poise.

$\therefore$ 1 poise $\equiv$ 1 dynes . sec cm^{-1}.

Other units in use are centipoise (0.01 *poise*) and *millipoise* (0.001 *poise*).

Therefore, the coefficient of viscosity (η) 'Can be defined' as the force in dynes that must be exerted between two parallel planes 1 cm^2 in area and 1 cm apart in order to maintain a laminar velocity difference of 1 cm sec of one layer past another.

METHODS OF DETERMINING VISCOSITY COEFFICIENT

The method commonly used for determining viscosity coefficient is based on Poiseuilles equation,

$$\eta = \frac{\pi P r^4 t}{8 L V} \quad \text{...(3)}$$

where V = volume (in cc) of a liquid of viscosity η that flows through a capillary tube of radius r cm and length L cm in time t secs under a pressure head of P dynes/cm^2.

Instead of determining η by a direct application of the Poiseuille equation, the one commonly used is the one devised by Ostwald in which the viscosity of one liquid is compared with that of a liquid of known viscosity. (*e.g.*, H_2O).

Use of Ostwald Viscometer

A known volume (say V_0 cc) of the given liquid is introduced into bulb B of the viscometer. The liquid is then gently sucked into the bulb A above the mark 'x' and the time (t_2) required for the liquid to flow through the capillary from 'x' to 'y' is accurately determined using a stop watch. A mean of three observations is recorded. The viscometer is then cleaned and dried and the same volume (V_0 cc) of water, of known viscosity, is introduced into bulb B. This is again sucked into bulb A gently and the time (t_1) required for water to flow between marks 'x' and 'y' is determined as before.

From equation (3), we have,

For water $$\eta_1 = \frac{\pi P_1 . r^4 . t_1}{8 L V}$$

For the given liquid $\eta_2 = \dfrac{\pi P_2 . r^4 . t_2}{8LV}$

Note that V = volume corresponding to that between the marks 'x' and 'y'

$$\therefore \quad \frac{\eta_2}{\eta_1} = \frac{P_2 . t_2}{P_1 . t_1} \quad ...(4)$$

But $\quad P = \dfrac{\text{Force}}{\text{Area}} = \dfrac{mg}{A} = \dfrac{V_0 . \rho g}{A}$

$$\therefore \quad \frac{P_2}{P_1} = \frac{V_0 . \rho_2 g/A}{V_0 \rho_1 g/A} = \frac{\rho_2}{\rho_1}$$

From equation (4), we get

$$\frac{\eta_2}{\eta_1} = \frac{\rho_2 . t_2}{\rho_1 . t_1}$$

$$\therefore \quad \eta_2 = \frac{t_2}{t_1} \times \frac{\rho_2}{\rho_1} \times \eta_1 \quad ...(5)$$

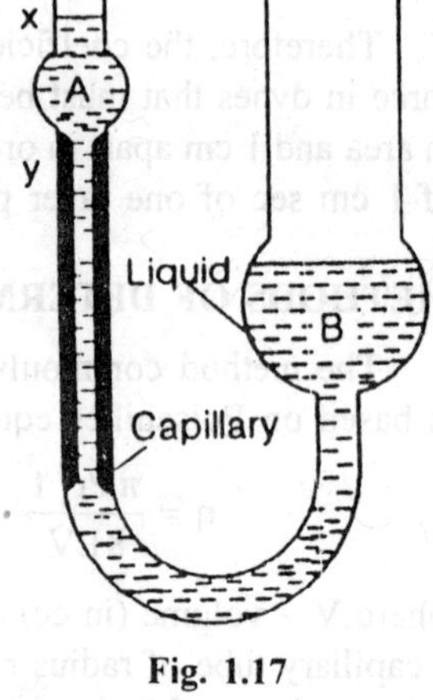

Fig. 1.17

Since all quantities on the R.H.S. of equation (5) are known, η_2 can be calculated. This is called the *absolute viscosity* of the given liquid. Then η_2/η_1 is the relative viscosity of liquid 2 w.r.t. liquid 1. If liquid 1 is water, then instead of using the term "relative viscosity", for the ratio η_2/η_1 we use the term "*specific viscosity*".

Determination of η by the 'Falling Sphere' Method

This method is specially useful for liquids having a high coefficient of viscosity.

Principle : A spherical body of a radius r and density p falling under gravity through a liquid medium of density ρ is acted upon by the gravitational force f_1 which is given by

$$f_1 = \frac{4}{3}\pi r^2(\rho - \rho')g \qquad ...(6)$$

This force which tends to accelerate the body falling through the liquid is opposed by the frictional forces within the liquid which increase as the velocity of the falling body increases. Ultimately, a uniform rate of fall is attained at which the frictional forces become equal to the gravitational force and thereafter the body continues to fall with a constant velocity v called the *terminal velocity*. Stokes showed that this frictional force f_2 is given by

$$f_2 = 6\pi r\eta v \qquad ...(7)$$

where r = radius of sphere

v = terminal velocity

η = coefficient of viscosity of the liquid.

Equating the two forces, we get

$$\frac{4}{3}\pi r^3(\rho - \rho')g = 6\pi r\eta \upsilon$$

$$\therefore \qquad \eta = \frac{2r^2(\rho - \rho')g}{9v} \qquad ...(8)$$

Procedure : The viscometer consists of a vertical tube filled with the given liquid and placed in a thermostat at a particular temperature. A steel ball of density ρ and of such diameter to give a slow rate of fall is then dropped through the neck of the tube. Its time of fall between the marks 'x' and 'y' is ascertained accurately by means of a stop-watch. Another liquid of known viscosity and density is now placed in the cylinder and the time required for the same steel ball to fall between the marks 'x' and 'y' is determined. Hence from equation (8) it follows that

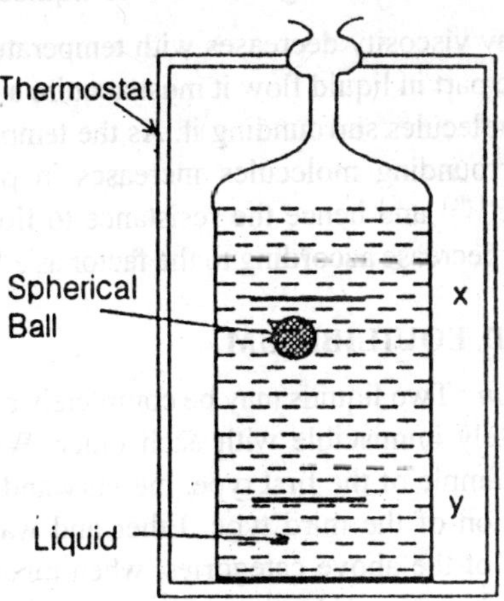

Fig. 1.18.

$$\frac{\eta_1}{\eta_2} = \frac{(\rho - \rho_1')}{(\rho - \rho_2')} \times \frac{v_2}{v_1}$$

But $$v = \frac{\text{distance between 'x' and 'y'}}{\text{time of fall}}$$

$$\therefore \quad \frac{\eta_1}{\eta_2} = \frac{(\rho - \rho_1')}{(\rho - \rho_2')} \times \frac{t_1}{t_2} \qquad ...(9)$$

EFFECT OF TEMPERATURE ON VISCOSITY

It has been found that the viscosity of a liquid decreases markedly with rise in temperature. The variation of viscosity with temperature is best expressed by the equation.

$$\eta = A.e^{E/RT} \qquad ...(1)$$

where A and E are constants for a given liquid.

Taking logarithms

$$ln\ \eta = ln\ A + \frac{E}{RT}$$

$$\therefore \log_{10} \eta = \log_{10} A + \frac{0.4343\ E}{RT} \qquad ...(2)$$

Hence, a plot of $\log_{10} \eta$ versus 1/T should be linear. This has actually been borne out for a large number of liquids.

The reason why viscosity decreases with temperature is that before a molecule can take part in liquid flow it must acquire a sufficient energy to push aside the molecules surrounding it. As the temperature increases the number of surrounding molecules increases in proportion to the Boltzmann factor $e^{-E/RT}$ and hence the resistance to flow *e.g.*, viscosity may be expected to decrease according to the factor as $e^{-E/RT}$ equation (1).

LIQUID—LIQUID EQUILIBRIUM

(a) *Introduction :* Two liquids may be completely miscible, partially or completely immiscible with each other. Water and alcohol form an example of the first type, mercury and water constitute an illustration of the third type. Ether and water belonging to the second of the above categories, when mixed together form to separate layers, one being an aqueous solution of ether and the other, ethereal solution of water. The two solutions thus formed are known as conjugate solutions. It is with such cases of partial miscibility of liquids that we shall be concerned, because they represent this type of equilibria faithfully. Complete miscibility yields only one phase, while we are going to consider the cases of two components in two phases and so they are left out from our considerations. Complete immiscibility, however, yields two phases no doubt, but each phase is constituted of the pure component (only one) and as such loses interest in this connection.

(b) *Classification of L-L System* : L-L systems can be classified into three different classes:

(i) The liquid pairs completely miscible in all proportions, *e.g.*, water-alcohol, benzene-nitrobenzene etc.

(ii) The liquid pairs partially miscible with one another, *e.g.*, phenol-water, nicotine-water etc.

(iii) The liquid pairs completely immiscible with one another, *e.g.*, water-benzene, alcohol- nitrobenzene etc.

Completely Miscible Liquid Pairs

There are certain pairs of liquids which dissolve in each other in all proportions. The general rule about solubility has been that closer the chemical similarity of the two liquids, the greater will be their miscibility. The expression, *"Similia Similibus Solvunter"* (like dissolved by like) is fully justified.

It is evident that the partial pressure of the components can be determined by the composition of the liquid and its temperature. This statement is present in *Duhem-Margales* equation for two components A and B, according to which

$$\frac{d \log P_A}{d \log x_A} = \frac{d \log P_B}{d \log x_B} \qquad ...(1)$$

The terms x and p denote the mole fraction and partial pressure of the respective components.

Derivation of Duhem-Margules Equation

Suppose there is a binary solution which is having n_B moles of A and n_A mole of B. Applying *Gibbs-Duhem Equation*, we get at equilibrium:

$$n_A d\mu_A + n_B d\mu_B = 0 \qquad ...(2)$$

were μ_A and μ_B denote the chemical potentials of A and B.

On dividing (2) by $(n_A + n_B)$, we obtain

$$\frac{n_A}{n_A + n_B}.d\mu_A + \frac{n_B}{n_A + n_B}.d\mu_B = 0 \quad ...(3)$$

where, x_A and x_B denote the mole fractions of constituents A and B respectively.

If the liquid mixture of A and B has been in equilibrium with their vapour phase at constant temperature T, the chemical potentials would be the same in both the phases for each component. Since,

$$\mu = RT \log f_i + \mu_i^0 \quad \text{(where } f_1 = \text{fugacity)}$$

or if the vapours are to behave ideally, we can write

$$\mu_i = RT \log p_i + \mu_i^0$$

where P denotes partial pressure of ith component in the vapour phase.

Thus $\mu = RT\ d \log p_i$.

On substituting it in equation (3), we get

$$x_A\ RT\ d \log p_A + x_B\ RT\ d \log p_B = 0$$

or
$$\frac{x_A\, d\log p_A}{d x_A} + \frac{x_B\, d\log p_B}{d x_B} = 0$$

$\because$
$$x_A + x_B = 1,\ dx_A = -dx_B.$$

$\therefore$
$$\frac{x_A\, d\log p_A}{d x_A} + \frac{x_B\, d\log p_B}{d x_B} = 0$$

$$\frac{x_A\, d\log p_A}{d x_A} + \frac{x_B\, d\log p_B}{d x_B}$$

or
$$\frac{d\log p_A}{d\log x_A} = \frac{d\log p_B}{d\log x_B} \quad ...(4)$$

Equation (4) is known as *Duhem-Margules Equation.* It gives rise to a correlation between the composition in the liquid phase and the partial vapour pressures in the gaseous phase.

Konowaloffs Rule

It is also possible to write the *Duhem-Margules Equation* (4) in the form

$$\frac{x_A}{p_A}.\frac{dp_A}{dx_A} = \frac{x_B}{p_B}.\frac{dp_B}{dx_B} \quad ...(5)$$

or since $dx_A = -dx_B$

$$\left[x_A = \frac{n_A}{n_A + n_B},\ x_B = \frac{n_B}{n_A + n_B}, \right.$$

where n denotes number of moles of the components

$$x_A + x_B = \frac{n_A}{n_A + n_B} + \frac{n_B}{n_A + n_B} = 1$$

or $\quad x_A = -x_B$

or $\quad \left. dx_A = -dx_B \right]$

Hence, (4) becomes as follows:

$$\frac{x_A}{p_A} \cdot \frac{dp_A}{dx_A} + \frac{x_B}{p_B} \cdot \frac{dp_B}{dx_B} = 0$$

Hence. $$\frac{dP}{dx_A} = \frac{dp_A}{dx_A} + \frac{dp_B}{dx_A}\left(1 - \frac{x_B \cdot p_A}{x_A \cdot p_B}\right), \qquad ...(6)$$

where P = total pressure.

The factor $\frac{dp_B}{dx_A}$ must be negative, because it is equal to $\frac{-dp_B}{dx_B}$

If $\frac{dp}{dx_A}$ is positive, then we have

$$x_B\ p_A > x_A\ p_B$$

or $$\frac{p_A}{p_B} > \frac{x_A}{x_B} \qquad ...(7)$$

If $\frac{dp}{dx_A}$ is negative *i.e.* $\frac{dp}{dx_B}$ is positive, then we have

$$x_B\ P_A < x_A\ p_B$$

or $$\frac{p_A}{p_B} < \frac{x_A}{x_B} \qquad ...(8)$$

Inequalities (7) and (8) have been the mathematical expressions of *Konowaloff's rule* that the vapour is richer in the component whose addition to the liquid mixture causes an increase of total vapour pressures.

Ideal Solutions of Liquids

An ideal solution may be defined as one in which the partial pressure of each component is proportional to the molar concentration in the mixture. This is also called *Raoult's law*. It may be mathematically stated in the form

$$p_A = p_A^0 \cdot x_A,$$

$$p_B = p_B^0 \cdot x_B,$$

where,

p_A, p_B = partial pressure of the respective components A and B

p_A^0, p_B^0 = vapour pressure of the two components

x_A, x_B = molar fraction of the two components.

Thus this law may be stated that the partial vapour pressure of any volatile constituent of a solution would be equal to its vapour pressure in the pure state multiplied by the mole fraction of two constituents in the solution.

Vapour Pressure—Composition Curve

Suppose there is a liquid pair A and B which are volatile and completely miscible and form an ideal solution, in which their mole fractions are x_A and x_B. Then their partial vapour pressures are as follows :

$$P_A = x_A . p_A^0$$

and $$p_B = x_B . p_B^0 \quad ...(9)$$

where, p_A^0 and p_B^0 denote their vapour pressures in pure state.

The total vapour pressure (P) of the solution may be given by

$$P = P_A + P_B = x_A\, p_A^0 + x_B p_B^0 \quad ...(10)$$

$$P = (1 - x_B)p_A^0 + x_B p_B^0 = P_A^0 + (p_B^0 - p_A^0)x_B \quad ...(11)$$

The relations (9), (10) and (11) have been such that if p_A, p_B or P are made to plot against mole fraction of any component, straight lines will be obtained which are shown in Fig. 1.19.

In the given Fig. 1.19, the dotted lines are giving plots of partial vapour pressures of A and B. It is also possible to derive the composition of the vapour by combining Dalton's law with Raoult's law. Let P_A = x_A'P. where x_A' denotes the composition of the vapour phase in equilibrium with the condensed phase. The following relationship can be derived for the vapour composition (only component A and B in system);

$$P = \frac{p_A^0 \cdot p_B^0}{p_A^0 - x_A^0\left(p_A^0 - p_B^0\right)} \qquad ...(12)$$

If the pressure of each constituent is plotted against its corresponding mole fraction in the liquid phase, then the plot would be a straight line which is passing through the origin.

By geometry, the plot of total vapour pressure would also be a straight line.

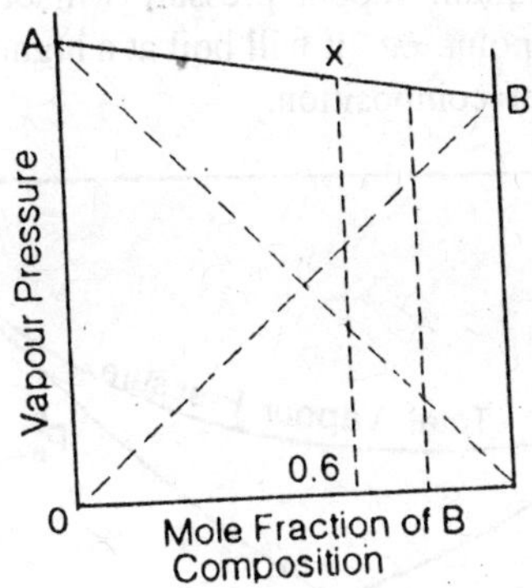

Fig. 1.19

In the given Fig. 19 A and B denote the vapour pressure of the pure components A and B. The line obtained by joining A and B would be the vapour pressure curve of the mixture of A and B of all proportions. Thus any point x on AB denotes the vapour pressure of the mixture say 0.6 mole fraction of B.

It is evident that the vapour pressure x would be given by:

$$y = 0.4 \times (\text{vapour pressure of pure A}) + 0.6 \times (\text{vapour pressure of pure B})$$

$$= (0.4 \times AC) + (0.6 \times BD).$$

The dotted lines BC and AD would give the partial pressures of the two components in the given various mixtures.

Ideal solutions get formed by chemically similar components, *e.g.*, ethyl bromide and ethyl iodide or n-hexane and n-heptane etc.

Real Solutions of Liquids

Only a few solutions of liquids are known to obey Raoult's law over the whole range of concentration, *i.e.*, in such a case plot of vapour pressure and composition of the liquid will yield a straight line. In most cases, either a positive or negative deviation can be seen.

(a) Type I : When Negative Deviations are Observed

In this case, the vapour pressure curve is having minimum as shown in Fig. 1.20, *e.g.*, pairs of acetone- chloroform, water-sulphuric acid. The solution with the minimum vapour pressure will obviously be having the maximum boiling point, *i.e.*, it will boil at a higher temperature than a mixture of any other composition.

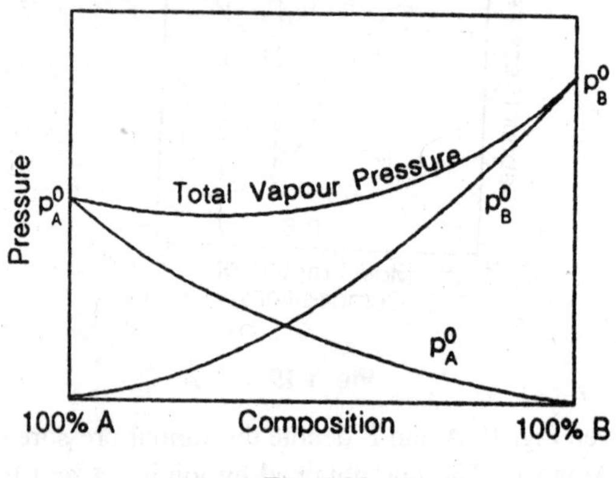

Fig. 1.20

Liquid pairs showing such behaviour have been as follows:

Toluene—acetic acid; acetone—chloroform; water—nitric acid; pyridine—acetic acid, etc.

Type II : When positive deviations are seen

Most mixtures are known to exhibit positive deviations from Raoult's law, *i.e.*, the measured vapour pressure would be greater than the ideal one. The shape of the curve is shown in Fig. 1.21.

In this case, the vapour pressure curve exhibits a maximum. The liquid pairs which are not chemically alike, undergo this type of behaviour, *e.g.*, acetone-carbon disulphide, chloroform-ethyl alcohol, benzene-toluene etc.

Large positive deviations occur in the case of mixtures of substances which are different in polarity internal pressure, length of hydrocarbon chain and degree of association. The solution having the maximum vapour pressure will be having the minimum boiling point, *i.e.*, it will boil at a lower temperature than a mixture of any other composition.

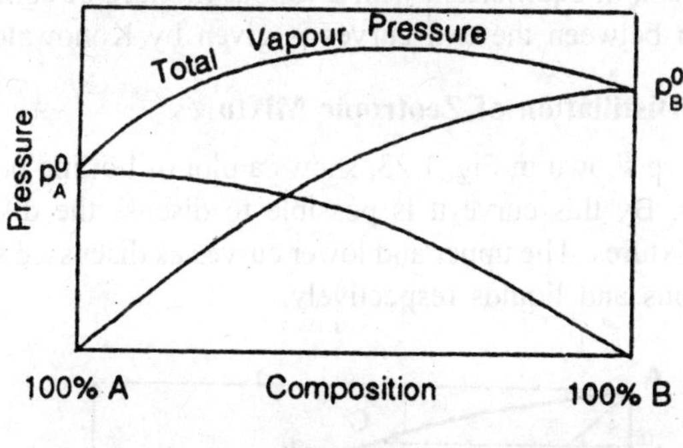

Fig. 1.21

Increase in temperature, however, decreases the extent of deviations from ideal behaviour, whereas a decrease in temperature will increase the deviations, until the mixture gets separated into two layers. It is possible to classify the completely miscible liquids into two types:

(I) Zeotropic mixtures.

(II) Azeotropic mixtures with a maximum or minimum boiling point.

Zeotropic Mixtures

Mixtures showing neither a maximum nor a minimum on the vapour pressure-composition curve are termed as zeotropic mixtures.

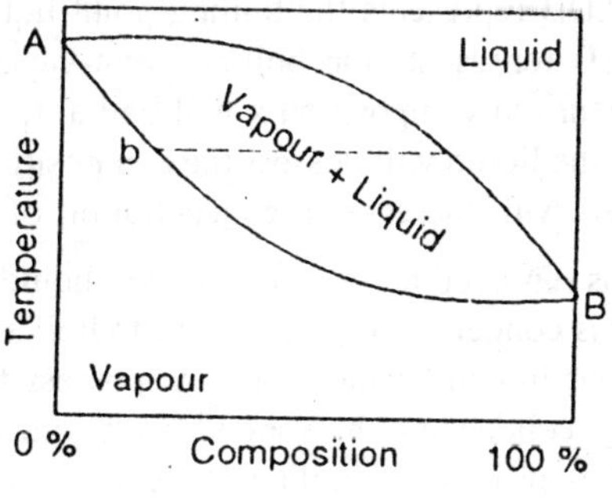

Fig. 1.22

In the given Fig. 1.22 the upper curve represents the liquids, which yields the composition of the liquids, while the lower represents the vaporous curve which yields the composition of the vapour in equilibrium with the liquid. In between the two curves, there exists a heterogeneous mixture of vapour and liquid. Hence, a liquid mixture of composition a would remain in equilibrium with a vapour mixture of composition b. The relation between the two curves is given by Konowaloff's rule.

Fractional Distillation of Zeotropic Mixtures

The curve shown in Fig. 1.23, shows a plot of boiling point against composition. By this curve it is possible to discuss the distillation of zeotropic mixtures. The upper and lower curves as discussed above have been vaporous and liquids respectively.

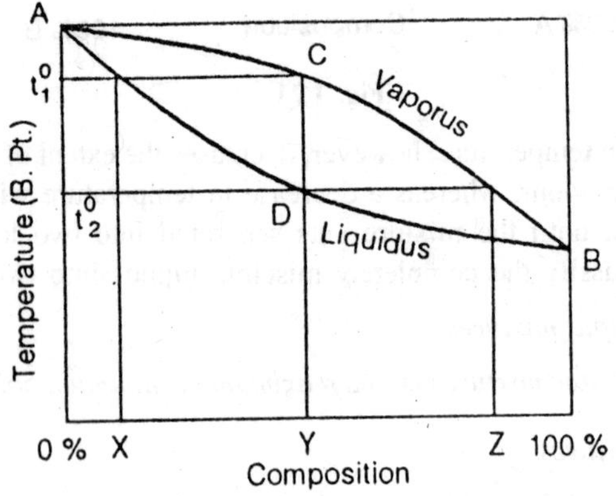

Fig. 1.23

In Fig. 1.23, ADB represents the boiling point-liquid composition curve, whereas ACB represents the boiling point-vapour composition curve. A liquid mixture of composition x will boil at t_1° and the vapour in equilibrium with the liquid will possess the composition y; the boiling point gets decreased with increase in concentration of z.

If this vapour is removed and condensed, the liquid will posses the composition y. If this condensed liquid is made to boil again, it will boil at t_2°, and the vapour in equilibrium will now possess the composition z. On removing and condensing the vapour, the liquid will possess the composition z. If this process of boiling and condensing in successive fractions is repeated, ultimately the pure component B would be obtained.

In the residual liquids, the component A would be gradually increasing. Thus, the two liquids can be separated from a mixture of this type.

It is possible to avoid the tedious process of successive evaporation and condensation by using a fractionating column in distillation. The process of separating two or more volatile liquids having different vapour pressures, from one another is termed as *fractional distillation.* The apparatus generally used in laboratories for fractional distillation is depicted in Fig. 1.23.

The apparatus is having a fairly long glass tube which is fitted vertically above the distillation flask and packed with glass beads, broken

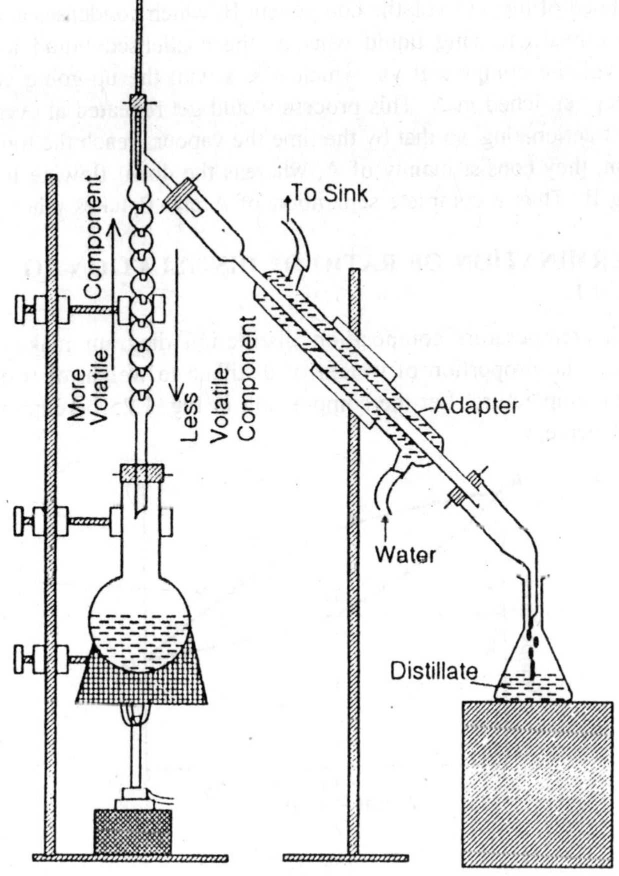

Fig. 24

glass tubes or blown into a series of spherical or pear shaped bulbs. The main aim is to increase the cooling surface and to furnish obstruction to the passage of ascending vapours or descending liquid. The liquid mixture is then made to heat. When it boils, the vapour rises in the column and progressively cools when it goes higher and higher. Suppose the mixture is having A which is more volatile component than B. When vapour comes in contact with large cooling surface of the fractionating column, the vapours of B get condensed whereas those of A pass over and collect in the receiver. As the condensed liquid, consisting of mostly less volatile component B, flows down the fractionating column, it meets the hot ascending vapours. During this process, the ascending vapours get robbed of the less volatile component B, which condenses and joins the downward flowing liquid whereas the condensed liquid loses its more volatile component y4, which mixes with the up-going vapours now very enriched in A. This process would get repeated at every step of the fractionating, so that by the time the vapours reach the top of the column, they consist mainly of A, whereas the down flowing liquid is having B. Thus a complete separation of A and B takes place.

DETERMINATION OF RATIO OF DISTILLATION TO RESIDUE

The temperature-composition distillation diagram makes us to calculate the proportion of weight of distillate to weight of residue at a given temperature. Let the composition in Fig. 1.25 be expressed in weight percent.

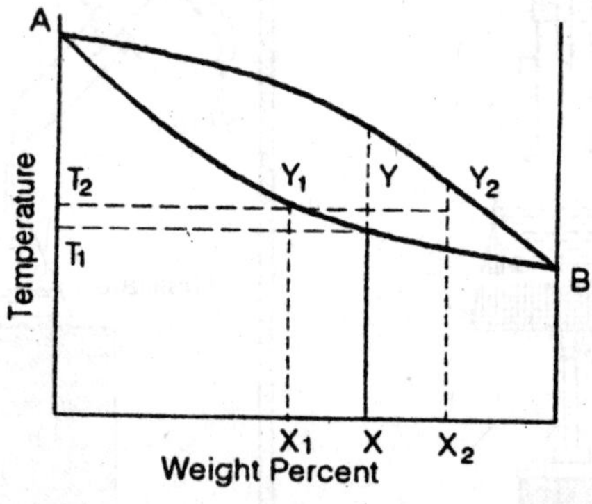

Fig. 1.25

Suppose we start with a liquid mixture of weight w having a composition of x% B. The liquid will start boiling only when the temperature gets raised to T_1°. With vaporisation, the liquid will get more concentrated in A and the boiling temperature will rise. At a temperature, say T_2°, the liquid will possess the composition as at y_1 which will be having x_1 % B and suppose the weight of the liquid now is w_1. If the system is closed, the vapour phase will be having the composition as at y_2, having x_2% B; the weight of the vapour phase is w_2.

Thus, $w = w_1 + w_2$, and the distribution of B would be such that

$$wx = w_1x_1 + w_2x_2$$

or $$(w_1 + w_2)\, x = w_1x_1 + w_2x_2$$

or $$w_1\,(x - x_1) = w_2\,(x_2 - x)$$

or $$\frac{w_1}{w_2} = \frac{x_2 - x}{x - x_1}$$

From the diagram, we can see

$$\frac{x_2 - x}{w - x_1} = \frac{yy_2}{yy_1}$$

Hence, $$\frac{\text{weight of distillate}}{\text{weight of residue}} = \frac{w_2}{w_1} = \frac{yy_1}{yy_2}$$

= ratio of the intercepts of the tie line y_1y_2.

Azeotropic Mixtures

A mixture which is possessing a lower vapour pressure (higher boiling point) or higher vapour pressure (lower boiling point) than any other mixture is termed as an azeotropic mixture. Such mixtures separate out on distillation into one of the components and a constant boiling mixture of both. Azeotropic mixtures are also termed as constant boiling mixtures possessing definite composition under a constant pressure.

If the vapour pressure-composition curve is having a minimum or maximum, then from equation (v), we have

$$\frac{dP}{dx} = 0.$$

Hence, $$x_Ap_A = x_Bp_B$$

or $$\frac{x_A}{x_B} = \frac{p_A}{p_B}$$

The other alternative $\frac{dp_B}{dx_A} = 0$ would be unlikely, because it would imply that the partial pressure would happen to remain constant in spite of a change in concentration. As the molecular ratio of the two components in the vapour may be given by $\frac{p_A}{p_B}$, if ideal behaviour is regarded, the composition of the vapour would be the same as that of the liquid with which it remains in equilibrium.

On applying phase rule in azeotropic mixtures, we should remember that there exists only one restriction viz; compositions in liquid and in vapour phase are the same.

Hence, $F' = C - P + 1 = 2 - 2 + 1 = 1$

i.e., azeotropes should behave as univariant systems, so that the boiling point will remain constant, if pressure is fixed. This is what is seen in the next two cases.

The characteristics of azeotropic mixtures are depicted in Figures 1.26 and 1.27.

(a) Distillation of Liquid Pairs with Maximum Boiling Point

The boiling temperature composition diagram is depicted in Fig. 26. The constant boiling mixture is having the maximum boiling point, *i.e.*, it is the least volatile. The vapour phase for any mixture lying between A and M would, therefore, get richer in A and mixture lying between B and M, would be richer in B than is the constant boiling mixture M.

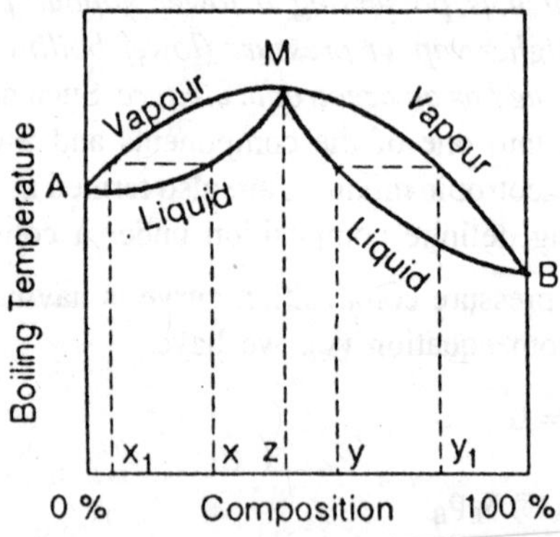

Fig. 1.26

Suppose, a mixture of composition x is make to distil. The first fraction distilled will be having composition indicated by x_1. Evidently, it is richer in A. The composition of the residual liquid, thus shifted towards constant boiling mixture M. As the distillation gets continued, the composition of the distillate gets changed towards A and that of the residue towards M. Ultimately, a distillate of pure A and a residue of constant boiling mixture M would be obtained.

Similarly, a mixture of composition lying between B and M, say y on distillation will ultimately provide a distillate of pure B and a residue of constant boiling mixture M.

It, therefore, follows that any binary solution of this type, on complete fractional distillation, can get separated into a residue of composition M and a distillate of either A and B depending upon whether the initial composition is lying between A and M at between B and M respectively. Thus, it becomes not possible to completely separate such a binary mixture into pure components A and B on distillation.

This mixture having the maximum boiling point is termed 'as *Maximum Boiling Azeotrope* and behaves as if it is a pure chemical compound of two components, as it boils at a constant temperature and the composition of the liquid and vapour is the same. But the azeotrope has been not a chemical compound, as its composition is not constant under all conditions and rarely corresponds to stoichiometric proportions.

Pure water and hydrogen chloride are made to boil at 100° and –85° while their constant boiling mixture (azeotropic mixture) having 20.25% of hydrogen chloride boils at 108.5°, under a pressure of 1 atmosphere. If a solution having less than 20.25% of HCl is made to distil (*i.e.*, between points A and M), water will pass over as the distillate and the residue left behind in the flask would be having 20.25% solution of HCl in water. Thus pure HCl cannot be obtained. Similarly, if a solution having more than 20.25% HCl is distilled, then pure HCl will pass over as distillate and the residue left behind in the flask will be having a mixture of the same constant composition, viz., 20.25% HCl in water.

(b) Distillation of Liquid Pairs with Minimum Boiling Point

If the distillation of composition represented by x is done then the first fraction collected will be having the composition x_1. It will be richer in the constant boiling mixture. The composition of the residual liquid will get shifted towards A. As distillation continues the composition of the distillate and residual liquid gets changed towards M and A respectively. By repeating this process, we get the mixture of minimum boiling point of composition M as distillate, while the residue left over in the distillation flask will be having only pure liquid A.

If we distil a liquid of composition represented by y, then the composition of the first fraction would be represented b y_1. Evidently, it will get richer in the constant boiling mixture. The composition of the liquid will gets richer in B. As the distillation carries out, the distillate and the residual liquid will get richer and richer in constant boiling mixture and pure B respectively. Finally, the distillate will be having only the constant boiling mixture and the residual liquid in the distillation flask will be having only B, There will occur no pure A in this case. *If the mixture is having the azeotropic composition (say z), it will distil unchanged.*

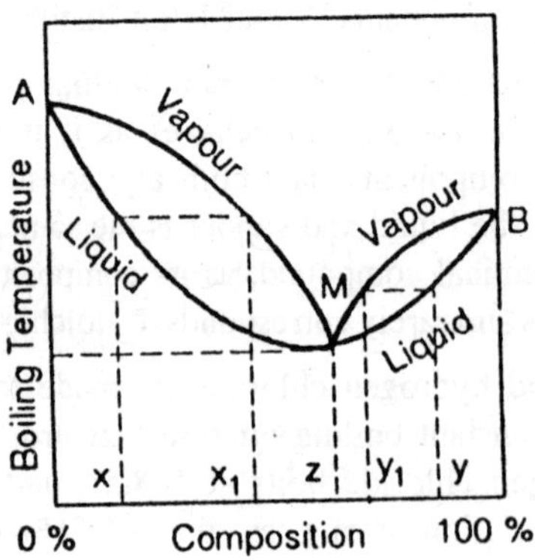

Fig. 1.27

In the system of water-ethanol the point M would correspond to a minimum boiling temperature of 78.13 and a composition of 95.57% ethanol by weight. If any solution of composition between pure water and 95.57% ethanol is made to distil, then we obtain a residue of pure water and a constant minimum boiling mixture of 95.57% alcohol in the distillate. No pure ethanol can be recovered. On the contrary, when a solution of composition between pure alcohol and 95.57% ethyl alcohol is made to distil, then we obtain mixture having 95.57% ethanol and pure alcohol. No pure water will yet recovered.

Binary Mixtures

(Minimum boiling point)

Components *A*	*B*	*Min. b. pt.*	*Composition % of B*
H_2O	Pyridine	92.6	59
H_2O	C_2H_5OH	78.13	95.57
C_6H_6	CH_3COOH	80.55	2.0
CS_2	Ethyl acetate	46.0	3.0

Partially Miscible Liquid Pairs

It is regarded that the actual relation between the partial pressures of the components would be given by

$$P_A = P_A^{\circ}.x_A.e^{\alpha}.x^2B$$

$$P_B = P_B^{\circ}.x_B.e^{\alpha}.x^2_A$$

where, a is a constant and all others are having their usual meaning.

If two liquids form a mixture and show a positive deviation from Raoult's law, the value of a is positive. On cooling the system, the positive deviation gets increased and then a also increases. It can be then by differentiating p by taking general case, with respect to x and equating it to zero. On rearranging, we obtained

$$2\alpha x^2 - 2\alpha x + 1 = 0$$

or $$x = \frac{2\alpha \pm \sqrt{4\alpha^2 - 8\alpha)}}{4\alpha}$$

The roots have been real if $(4\alpha^2 - 8\alpha)$ has been positive, *i.e.*,

$$4\alpha^2 - 8\alpha > 0,$$

or $$\alpha - 2 > 2,$$

or $$\alpha > 2.$$

The roots will be equal if $\alpha = 2$.

The roots would be imaginary when $\alpha < 2$. The corresponding curves are depicted in the given

The curve, a, b and c would correspond to increasing temperatures. At the highest temperature, the deviation is appreciably small (curve c),.

At the lowest temperature, curve a exhibits a maximum and minimum. The S shaped curve reveals that the three different solutions separated by x, y and z should be having the same vapour pressure.

Hence, only two liquid phase would be in equilibrium with vapour. Hence, three liquid phases cannot exist and in particle the liquid would break up into two liquid layers of composition x and y.

If the temperature is raised, points x and z come closer and closer, till at a certain temperature, the two points merge into one. In that case α = 2. This particular temperature is termed as *upper consulate temperature* of upper critical solution temperature (U.C.S.T.). At this temperature, the two liquid phases would be becoming identical. Above this temperature, the liquids will get miscible in all proportions. At the consulate temperature, the system would be non-variant

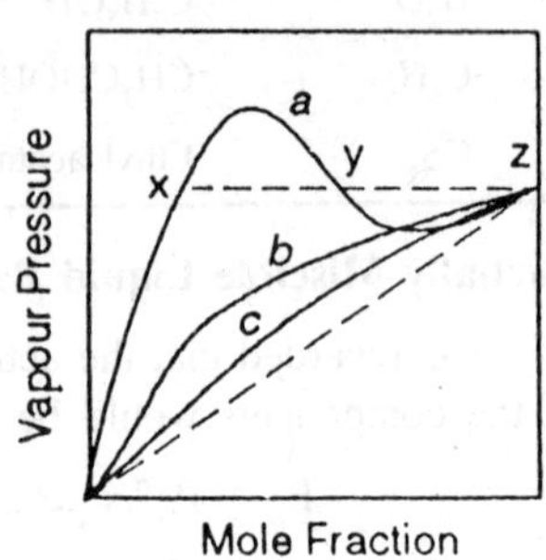

Fig. 1.28

So, if pressure is maintained constant temperature will be a fixed point.

Partially Miscible Liquid Pairs

There are quite a number of liquid pairs which show only *limited miscibility* in each other *i.e.*, they are only partially miscible. This is similar to the solution of a sparingly soluble salt in a liquid. For example, when a small amount of phenol or aniline is added to water at room temperature and shaken, a solution of phenol or aniline in water is obtained. However, on adding a larger amount of phenol or aniline, two liquid layers separate. One is a solution of phenol or aniline in water and the other a solution of water in phenol or aniline. *Each layer is a saturated solution of one in the other*. At constant temperature, the composition of the two layers, although different from each other, remain constant as long as the two phases are present. The two layers in equilibrium are called *conjugate solutions*. The addition of any one of the liquids merely changes the relative volumes of the two layers and not their composition. Further addition of phenol or aniline causes the water layer to diminish in size which finally disappears and only a layer of water in phenol or aniline remains. The mutual solubility of the two liquids varies in a characteristic manner with temperature. In this case, the mutual solubility goes on increasing as the temperature is increased until a homogeneous solution is obtained. *This temperature at which the two liquids become completely miscible is called the mutual solubility temperature (M.S.T.) for that particular composition* of the two liquids.

The variation of mutual solubility of phenol and water with temperature is as shown in Fig. 29. At any one temperature in the curve, the composition of each layers is fixed by points A and B.

A gives the composition of phenol in water while B gives that of water in phenol at that temperature. *The line joining the two points A and B is called the tie line.* With the rise of temperature as is seen from the curve, the compositions of the two layers approach each other and at 339K, the two layers merge into one homogeneous solution. This temperature is called the *critical solution temperature* (C.S.T.) or the *consulate temperature* of the system. Above this temperature, it is seen that the two liquids are miscible with each other in all proportions and only a single layer of liquid results.

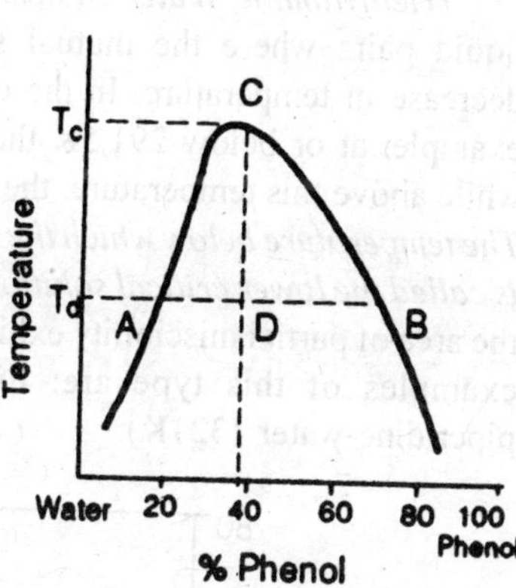

Fig. 1.29 : Mutual solubility diagram of the mixture showing upper C.S.T.

Since this temperature is the maximum on the mutual *solubility temperature composition curve*, it is termed the *upper critical solution temperature*.

Any solution of phenol and water of composition and temperature represented by a point within the curve ACB consists of two layers. The composition of the two layers is given by a horizontal line (parallel to x-axis) through the point cutting the curve at A and B. Any composition at a given temperature represented by points on the left of the curve A C or the right of the curve CB consists of only one layer. All compositions between pure phenol and point B yield a solution of water in phenol while all compositions between pure water and point A yield a solution of phenol in water. Thus it follows, within the dome shaped area ABC *i.e.*, the system is heterogeneous and two liquid phases exist, while in the area outside the dome only a single liquid layer *i.e.*, a homogeneous system exist. *The upper critical solution temperature may, therefore be defined as the temperature above which the two liquids which had been partially miscible, become miscible in all proportions.*

Other systems showing similar behaviour are aniline-water, (441K), aniline-hexane (332.6K) cyclohexane-methanol (322.1K), methanol-carbon disulphide (313.5K).

Triethylamine Water System : There are a few partially miscible liquid pairs where the mutual solubility is found to increase with decrease in temperature. In the case of *triethylamine-water system* for example, at or below 291.5K the two liquids are completely miscible, while above this temperature, the two liquids are only partially miscible. *The temperature below which the two liquids become completely miscible is called the lower critical solution temperature* since the curve confining the area of partial miscibility exhibits a minimum (Fig. 1.30). Some other examples of this type are: diethylamine-water (416K), 1-methyl piperidine-water (321K).

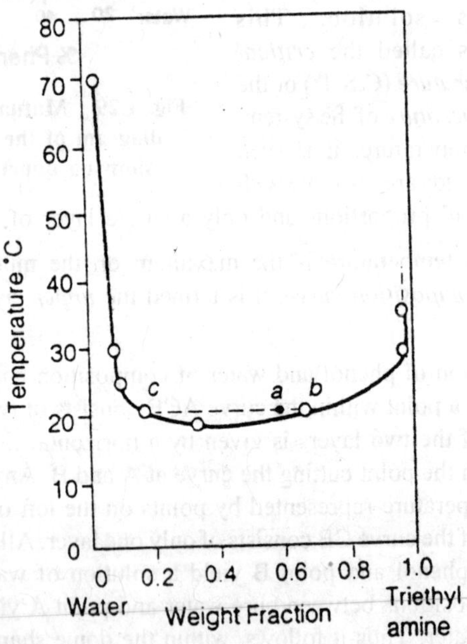

Fig. 1.30 : Solubility of triethylamine in water at various temperatures.

Nicotine Water System : This system exhibits an upper as well as lower critical solution temperature. Within the enclosed area, the two liquids are only partially miscible and a heterogeneous system exists while outside this area, there is only a single layer exists *i.e.*, a homogeneous phase is present. The upper or maximum CST is 481K. C in Fig. 1.31 while the lower or minimum CST is 333.8 K indicated by C'. The compositions corresponding to C and C' are the same equal to 34% nicotine.

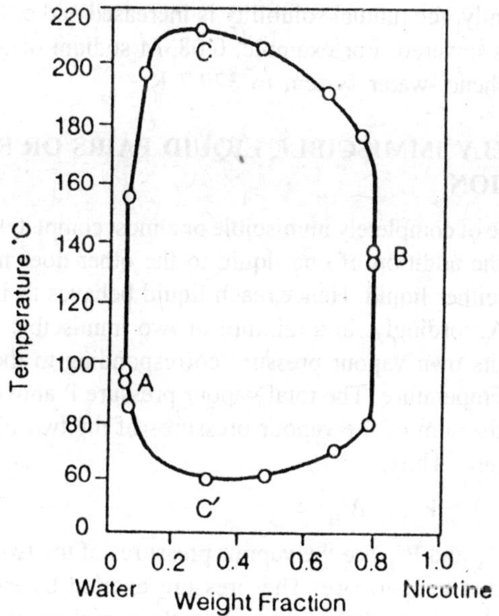

Fig. 1.31 : Solubility of nicotine in water at various temperatures.

The CST of this system is affected by pressure. On applying external pressure to the system, the upper and the lower CST approach each other until a pressure is reached when the two liquids become completely miscible. Other systems of this type are :

glycerol-w-toluidine (280K & 393K)

methyl ethyl ketone-water (279K & 406K)

It is believed that all partially miscible systems in general show an upper as well as a lower critical solution temperature. In many cases one of them may not be experimentally realized due to certain physical conditions.

There are also certain liquid pairs like diethyl ether-water which do not show an upper or a lower CST. They are only partially miscible in each other at all temperatures.

The critical solution temperature is affected considerably by the presence of foreign substances. If the foreign substance is soluble in only one of the liquids, the mutual solubility is decreased resulting in an increase in the critical solution temperature. For example, 0.15M KCl raises the critical solution temperature of phenol-water system by about 12K. On the other hand, if the foreign substance dissolves in both the liquids uniformly, the mutual solubility is increased and critical solution temperature is lowered. For example, 0.083M sodium oleate decreases the CST of phenol-water system to 329.7 K.

COMPLETELY IMMISCIBLE LIQUID PAIRS OR STEAM DISTILLATION

In the case of completely immiscible or almost completely immiscible liquid pairs, the addition of one liquid to the other does not affect the properties of either liquid. Hence, each liquid behaves as if the other is not present. Accordingly, in a mixture of two immiscible liquids, each liquid exerts its own vapour pressure corresponding to the pure liquid at the given temperature. The total vapour pressure P above the mixture is therefore, the sum of the vapour pressures of the two pure liquids at that temperature. Thus,

$$P = P^\circ_A + P^\circ_B$$

Where P°_A and P°_B are the vapour pressures of the two pure liquids A and B at that temperature. The pressure exerted by each layer and hence the total pressure do not depend on the actual or relative amounts of the liquids present.

Any system, as we know boils at a temperature when its total vapour pressure becomes equal to the external pressure. In this case, since the two liquids together can reach any given total pressure at a lower temperature than either liquid alone, it is obvious that the mixture would boil at a temperature than either liquid alone, it is obvious that the mixture would boil at a temperature lower than the boiling point of either of the two liquids. Further, since at any given temperature there is no change in total vapour pressure with change in composition, the boiling point of all possible compositions of any two immiscible liquids remains constant, as long as the two liquids are present. The temperature rises to T_A or T_B depending upon whether A or B remains, only when one of the liquids is boiled away.

The relative proportions of the two liquids in the distillate can be calculated, since the number of moles of each component in the vapour phase is proportional to its vapour pressure. At the boiling point T, if n_A and n_B are the number of moles of the two liquids A and 5 in the vapour phase, then

$$n_A \propto P^\circ_A$$

$$n_B \propto P^\circ_B$$

or
$$\frac{n_A}{n_B} = \frac{P^\circ_A}{P^\circ_B}$$

The composition of the vapour remains a constant, since P°_A and P°_B are constants at a given temperature T. If W_A and W_B are the actual weights of the two liquids A and B in the distillate, and M_A and M_B the respective molecular weights, then

$$\frac{P^\circ_A}{P^\circ_B} = \frac{W_A M_B}{W_B M_A}$$

or
$$\frac{W_A}{W_B} = \frac{P^\circ_A M_A}{P^\circ_B M_B}$$

The above equation relates directly the ratio of the weights of the two components present in the distillate of a mixture of two immiscible liquids to the molecular weights and vapour pressure of the two pure components.

Examples of such pairs include water-cyclohexane, water-nitrobenzene, water-bromobenzene etc., Distillation of immiscible liquids is utilized industrially and in the laboratory as it involves lowering of boiling points of the components. Purification of organic liquids which either have very high boiling point or tend to decompose when heated to their boiling point can be suitably carried out using the above principle. The other liquid is generally water and the process is referred to as *Steam Distillation.*

The immiscible mixture of the liquid and water is either heated directly or by passing the vapours of steam into the liquid, and the vapours distilling over are condensed and separated. In this manner it is possible to distil many organic liquids of high boiling point at temperatures below 373K *i.e.*, the boiling point of water.

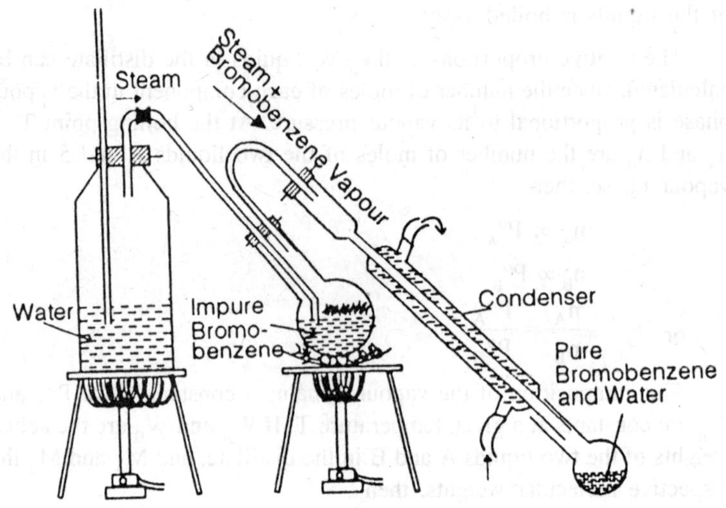

Fig. 1.32 : Steam distillation.

The apparatus is as shown in Fig. 1.32. Steam from the steam generator is passed into the round-bottom flask containing the liquid, say nitrobenzene or bromobenzene to be steam distilled. The tube carrying the steams dips into the liquid and the flask is heated gently on a sand bath to avoid too much condensation of water into it.

The vapours of the organic liquid mixed with steam distils over, condenses and collects in the receiver. The organic liquid is then separated from the aqueous layer and finally dried. As discussed earlier, the proportion by mass of the organic liquid that distils over is related directly to its vapour pressure and molecular weight. Thus a higher proportion of the liquid is obtained for liquids with high molecular weights and having a relatively high vapour pressure at about the boiling point of water.

SOLVED EXAMPLES

Example 1:

At 20°C, water formed 29 drops when flowing through the capillary of a stalagmometer, while an equal volume of ether formed 86 drops. If the densities of water and ether are 0.997 and 0.70 gm per cc respectively, find the surface tension of ether if that of water is 72.8 dynes/cm.

Solution:

We have $\gamma_2 = \frac{n_1}{n_2} \cdot \frac{\rho_2}{\rho_1} . \gamma_1$

$$= \frac{29}{86} \times \frac{0.07}{0.997} \times 718 = 17.24 \text{ dynes/cm}$$

Example 2:

At 293 K, $10^{-2} dm^3$ of water formed 29 drop sand the same volume of an organic liquid formed 86 drops in the same stalagmometer. Density of the organic liquid is 0.7 g cm^{-3}. If the surface tension of water is 7.2×10^{-2} N m^{-1}, what would be the surface tension of the organic liquid ?

Solution:

If n_1, ρ, and γ_1 are the number of drops, density and the surface tension of the organic liquid, and n_2, d_2 and γ_2 are the number of drops, density and the surface tension of water respectively, then we have

$$\frac{\gamma_1}{\gamma_2} = \frac{n_1\rho_1}{n_1\rho_2}$$

or $$\gamma_1 = \frac{\gamma_2 n_2 \rho_1}{n_1 \rho_2}$$

Since $n_2 = 29$ $\qquad n_1 = 86$

$\rho_2 = 1$ g cm^{-3} $\qquad \rho_1 = 0.70$ g cm^{-3}

$= 1 \times 10^3 gm^{-3}$ $\qquad - 7.0 \times 10^2$ gm^{-3}

$\gamma_2 = 7.2 \times 100^{-2} Nm^{-1}$ $\qquad \gamma_1 = ?$

Therefore

$$\gamma_1 = \frac{29(7.0 \times 10^2 \text{ kg m}^{-3})(7.2 \times 10^{-2} \text{ N m}^{-1})}{86 (10^3 \text{ kg m}^{-3})}$$

$= 1.726 \times 10^{-2}$ m^{-1}

or $= 17.26$ dynes cm^{-1}

Example 3:

Equal volumes of an organic liquid and water gave 55 drops and 35 drops respectively. The densities of water and the organic liquid are

0.996 and 0.80g cm^{-3} and the surface tension of water is 7.2 × 10^{-2} Nm^{-1}. Calculate the surface tension of the organic liquid. How many times a water drop is heavier than a drop of the organic liquid ?

Solution:

$$\frac{\gamma_1}{\gamma_2} = \frac{n_2 d_1}{n_1 d_2}$$

or
$$\gamma_1 = \frac{\gamma_1 n_2 d_1}{n_1 d_1}$$

Since

$n_1 = 55$ $\quad$ $n_2 = 35$

$d_1 = 0.80\text{g cm}^{-3}$ $\quad$ $d_2 = 0.996\text{g cm}^{-3}$

$= 8.0 \times 10^2 \text{ kg m}^{-3}$ $\quad$ $= 9.96 \times 10^2 \text{ kg m}^{-3}$

$\gamma_1 = ?$ $\quad$ $\gamma_2 = 7.2 \times 10^{-2} \text{ Nm}^{-1}$

Therefore

$$\gamma_1 = \frac{(7.2 \times 10^{-2}\text{ Nm}^{-1})(35)\,(8.0 \times 10^2\text{ kg m}^{-3})}{(55)\,(9.96 \times \text{kg m}^{-3})}$$

$$= 2.63 \times 10^{-2} \text{ Nm}^{-1}$$

Now,
$$\frac{m_{water}}{m_{organic\ liquid}} = \frac{\gamma_{water}}{\gamma_{organic\ liquid}}$$

$$= \frac{7.2 \times 10^2\text{ Nm}^{-1}}{2.63 \times 10^{-2}\text{ Nm}^{-1}} = 2.74.$$

Example 4:

Water requires 120.5 seconds to flow through a viscometer and the same volume of acetone requires 49.5 seconds. If the densities of water and acetone at 293 K are 9.982 × 10^2 kg m^{-3} and 7.92 × 10^2 kgm^{-3} respectively and the viscosity of water at 293 K is 10.05 pascal second, calculate the viscosity of acetone at 293 K.

Solution:

Since,
$$\frac{\eta_{water}}{\eta_{Acetone}} = \frac{t_1 \rho_1}{t_2 \rho_2}$$

$$\eta_{Acetone} = \frac{\eta_{water}\, t_2 \rho_2}{t_1 \rho_2}$$

$$= \frac{(1.005 \times 10^1 \text{ Pas})(49.5\text{s})(7.92 \times 10^2 \text{ kg m}^{-3})}{(12.5\text{s})(9.982 \times 10^2 \text{ kg m}^{-3})}$$

Example 5:

The coefficients of viscosity of two liquids at 298 K are 1.408 × 10^{-3} kg m^{-1} s^{-1} and 1.594 × 10^{-3} kg m^{-1} s^{-1} and their densities at the same temperature are 8.07 × 10^{-3} kg m^{-3} and 10.17 × 10^2kg m^{-3} respectively. If the time of flow in an Ostwald viscometer for the first liquid is 100 seconds, calculate the time of flow for the second liquid.

Solution:

$$\frac{\eta_1}{\eta_2} = \frac{t_1\rho_1}{t_2\rho_2}$$

or $$t_2 \frac{\eta_2}{\eta_1} = \frac{t_1\rho_1}{\rho_2}$$

$$= \frac{(1.594 \times 10^{-3} \text{ kg m}^{-1}\text{s}^{-1})(100\text{s})(8.07 \times 10^2 \text{ kg m}^{-3})}{(1.408 \times 10^{-3} \text{ kg m}^{-1}\text{ s}^{-1})(10.7 \times 10^2 \text{ kg m}^{-3})}$$

Example 6:

A steel ball of density 7.90 gm per cc and 4 mm diameter requires 55 sees to fall a distance of 1 meter through a liquid of density 1.10 gm per cc. Calculate the viscosity of the liquid in poises.

Solution :

According to equation (8), we have

$$\eta - \frac{2r^2 (\rho - \rho')g}{9v}$$

where, r = Radius = 0.2 cm

v = Velocity = 100/55 cm sec^{-1}

$$\eta = \frac{2 \times (0.2)^2 (7.90 - 1.10) \times 980}{9 \times \frac{100}{55}} = 32.58 \text{ poises}$$

Example 7:

A sphere of 5 × 10^{-2} cm and density 1.10 gm per cc falls at constant velocity through a liquid of density 1.00 gm per cc and viscosity 1.00 poise. What is the velocity of the falling sphere?

Solution:

Rearranging equation (8) we have

$$v = \frac{2r^2 (\rho - \rho')g}{9\eta}$$

On substituting

$$v = \frac{2 \times (5 \times 10^{-2})^2 \times (1.10 - 1.00) \times 980}{9 \times 1.00}$$

$$= 5.434 \times 10^{-3} \text{ cm/sec.}$$

Example 8:

Two immiscible liquids, water and chlorobenzene are boiled at 97.89 kNm⁻² pressure and the mixture boils at 363K. The ratio of the weight of chlorobenzene to water collected in the distillate is 2.47. Calculate the molecular weight of chlorobenzene if vapour pressure of water at this temperature is 70.11 kNm⁻².

Solution:

Vapour pressure of chlorobenzene

$$= 97.89 \text{ kNm}^{-2} - 70.11 \text{ kNm}^{-2} = 27.78 \text{ kNm}^{-2}$$

$$\frac{\text{Weight of chlorobenzene}}{\text{Weight of water}} = \frac{P^\circ_A \, M_A}{P^\circ_B \, M_B}$$

i.e.

$$2.47 = \frac{27.78 \text{ kNm}^{-2} \times M_A}{70.11 \text{ kNm}^{-2} \times 0.018 \text{ kgmol}^{-1}}$$

∴ M_A *i.e.*, molecular weight of chlorobenzene

$$= \frac{2.47 \times 70.11 \text{ kNm}^{-2} \times 0.018 \text{ kgmol}^{-1}}{27.78 \text{ kNm}^{-2}} = 0.1122 \text{ kgmol}^{-1}$$

$$\text{Relative molar mass} = \frac{0.1122 \text{ kgmol}^{-1}}{0.001 \text{ kgmol}^{-1}} = 112.2.$$

Example 9:

The radius of a given capillary is 1.05 × 10⁻⁴ m. A liquid whose density is 0.80g/cm³ rises in the capillary up to a height of 6.25 × 10⁻²m. Calculate the surface tension of the liquid assuming the contact angle θ to be zero.

Solution:

Here $r = 1.05 \times 10^{-4}$m $\qquad d = 0.80$g/cm³

$h = 6.25 \times 10^{-2}m \quad = 8.0 \times 10^{2}kg/m^{3}$

$g = 9.8ms^{-2}$

Since, $\gamma = 1/2\ rhdg$

$\therefore \gamma = \frac{1}{2}(1.05\times10^{-4}m)\ (6.25\times10^{-2}m)\ (8.0\times10^{2}kg\ m^{-3})(9.8ms^{-2})$

$= 2.5725 \times 10^{-2}kg\ s^{-2}$

$= 2.5725 \times 10^{-2}Nm^{-1}$ $(1\ N = 1kg\ ms^{-2})$

Example 10:

The surface tension of toluene at 293K is 0.0284 Nm^{-1} and its density at this temperature is 0.866g cm^{-3}. What is the largest radius of the capillary that will permit the liquid to rise 2×10^{-2} m ?

Solution:

Since $\gamma = \frac{1}{2}$ rhdg

Therefore $r = \frac{2r}{hdg}$

Here $\gamma = 2 \times 10^{-2}$ m $\quad h = 0.866\ g\ cm^{-3}$

$\gamma = ?$ $\quad = 8.66 \times 10^{2}\ kg\ m^{-3}$

$\gamma = 0.0284\ N\ m^{-1}$ $\quad g = 9.8\ m\ s^{-2}$

Substituting these values in the above equation, we get

$$r = \frac{2(0.084\ N\,m^{-1})}{(2\times10^{-2}\ m)\ (8.66\times1^{2}\ kg\,m^{-3})\ (9.8\ ms^{-2})}$$

$= 3.347 \times 10^{-4}\ kg^{-1}\ s^{2}m^{2}$

$= 3.347 \times 10^{-4}m$ $\quad (1N - 1\ kg\ ms^{-2})$

$= 0.03347$ cm

Example 11:

At a pressure of 101.3kNm^{-2} a mixture of water and nitrobenzene distils over at 372K. The vapour pressure of water at 372K is 97.70 kN m^{-2}. Estimate the proportion by weight of nitrobenzene in the distillate.

Solution:

$$\frac{\text{Weight of nitrobenzene } W_A}{\text{Weight of water } W_B} = \frac{P^{\circ}_A\ M_A}{P^{\circ}_B\ M_B}$$

Molecular weight of nitrobenzene = 0.123 $kgmol^{-1}$

Molecular weight of water = 0.018kg mol^{-1}

$P°_B = P - P°_A = 101.3 \text{ k Nm}^{-2} - 97.70 \text{ kNm}^{-2} = 3.60 \text{ kNm}^{-2}$

$$\therefore \quad \frac{W_A}{W_B} = \frac{3.60 \text{ kNm}^{-2}}{97.70 \text{ kNm}^{-2}} \times \frac{0.123 \text{ gmol}^{-1}}{0.018 \text{ kgmol}^{-1}}$$

∴ WA : WB = 0.25 : 1.00

∴ Proportion of nitrobenzene in the distillate

$$= \frac{0.25}{1.25} \times 100 = 20\%.$$

Example 12:

In a steam distillation of an insoluble oil, the mixture boils at 368K when the external pressure is 100 kNm^{-2}. If the vapour pressure of water at this temperature is 84.0 kNm^{-2}. If the molecular weights of water and oil are 0.018 kg mol^{-1} and 0.380 kg mol^{-1} respectively, calculate the amount of steam required to distil 0.1 kg of oil.

Solution:

$$\frac{\text{Weight of steam } W_A}{\text{Weight of oil } W_B} = \frac{P°_A \, M_A}{P°_B \, M_B}$$

$$P°_A = 84.0 \text{ kNm}^{-2}$$

$$P°_B = 100 - 84.0 = 16 \text{ kNm}^{-2}$$

$$\frac{W_A}{W_B} = \frac{84.0 \text{ kNm}^{-2}}{16.0 \text{ kNm}^{-2}} \times \frac{0.018 \text{ kg mol}^{-1}}{0.380 \text{ kgmol}^{-1}}$$

Weight of oil in the distillate = 0.1 kg.

$$\therefore W_A = \frac{84.0 \text{ kNm}^{-2}}{16.0 \text{ kNm}^{-2}} \times \frac{0.018 \text{ kg mol}^{-1}}{0.380 \text{ kgmol}^{-1}} \times 0.1 \text{ kg}$$

$$= 0.0248 \text{ kg}$$

i.e., weight of steam necessary to distil 0.1 kg of oil is 0.0248 kg.

2

SOLIDS STATE

INTRODUCTION

The classification of matter into solids, liquids and gases in termed physical classification of matter. Solid, liquids and gas are the three states of matter. Thus, matter exist in three physical stares; gas, liquid and solid. Solids under ordinary conditions have definite volume and shape and tend to maintain these even under deforming forces. Liquids also have definite volume but take the shape of the vessel into while they are poured. Both the liquids and solids are nearly incompressible. Gases, on the otherhand, maintain neither the volume nor shape and completely fill the container into which they are introduced. Gases can be expanded or compressed very easily.

The three states of matter are interconvertible. This can be done by heating or cooling. Heating increases the interparticle spacing and the kinetic energy of the particles. So a solid on interparticle spacing and the kinetic energy of the particles in shown in Fig. 2.1.

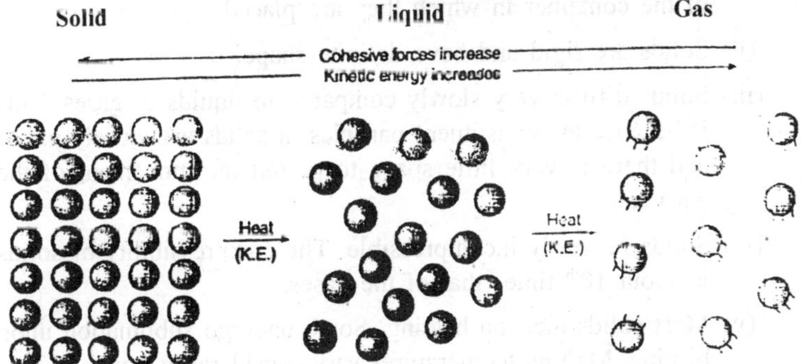

Fig. 2.1 : Effect off heating on the interparticle spacing and the physical state of the matter.

Many properties of solids, liquids and gases can be easily observed with the help of our sense organs. The properties which can be observed with the help of our sense organs are called macroscopic properties. The description of the behaviour of the three states of matter an terms of atomic theory is called microscopic description of matter. In chemistry, we try to explain the macroscopic behaviour of matter in terms of its microscopic description. In this chapter, we would try to explain the observable properties of different states of matter in terms of the behaviour of the constituent particles in term.

In a solid, the constituent particles atms, ions or molecules are most closely packed and have the strongest inter molecular force of attraction. As a result particles are fixed in their position and do not have any freedom of motion except that of vibration about their mean position. This give rise to the following characteristic properties to a solid.

(i) A definite shape and definite volume

(ii) Incompressibility

(iii) Rigidity

(iv) No fluidity

(v) Poor diffusibility through them

CHARACTERISTICS OF SOLIDS

Some characteristic properties of solids are given as follows :

(i) Solids maintain their volume independent of the size of shape of the container in which they are placed.

(ii) Solids are rigid and have definite shape.

(iii) Solids diffuse very slowly compared to liquids or gases. This is because the constituent particles in solids are closely paced and there is very little space to permit the movement of the particles.

(iv) Solids re nearly incompressible. The compressibility of solids is about 10th times that of the gases.

(v) Most solids melt on heating. Some undergo sublimation upon beating. Melting temperatures vary considerably from solids to solid.

(vi) Solids have very high mass-to-volume ratio (density) as compared to gases.

CLASSIFICATION OF SOLIDS

Based on their structural features, the solids are classified into following three type :

(a) Crystalline solids

(b) Polycrystalline solids

(c) Amorphous solids.

These are described below.

Crystalline Solids

All crystalline solids have the following characteristics.

(i) The constituent particles in crystals are generally held by strong interatomic interionic or intermolecular forces. Particles in a solid do not possess translation motion, but they vibrate about their equilibrium positions.

(ii) Crystalline solids possess characteristic geometrical shapes.

(iii) When cut or hammered gently, crystalline solids show a clean cleavage, *i.e.*, they show fracture along a smooth surface.

(iv) Crystalline solids have sharp melting points. This indicate that there exists a long range order in crystalline solids. The long range order in crystalline solids is due to regular arrangement of the constituent particles (atoms, ions or molecules) in three dimensions.

For example in a crystal of sodium chloride, experiments (X-ray diffraction method) show that Na^+ and Cl^- ions are located at alternate sites as shown below:

Na^+ Cl^- Na^+ Cl^- Na^+ Cl^-

Cl^- Na^+ Cl^- Na^+ Cl^- Na^+

Na^+ Cl^- Na^+ Cl^- Na^+ Cl^-

Cl^- Na^+ Cl^- Na^+ Cl^- Na^+

The arrangement shown above is only in two dimensions. In three dimensions we find that each Na^+ ion is surrounded by a fixed number (six) of Cl^- ions and *vice-versa*. The origin of the three dimensional order in NaCl crystals is due to the strong Colombian attraction between Na^+ and Cl^- ions. A similar regular arrangement is also found in other solids.

(v) The mechanical, electrical and optical properties depend upon the direction along which they are measured, *i.e.*, crystalline solids are anisotropic in nature.

Diamond is a crystalline slid. When hammered gently, diamond get cleaved smoothly along certain preferred direction.

Hammer gently a big crystal of rock salt (sodium chloride). The big crystal gets broken down into smaller crystals with smooth surfaces.

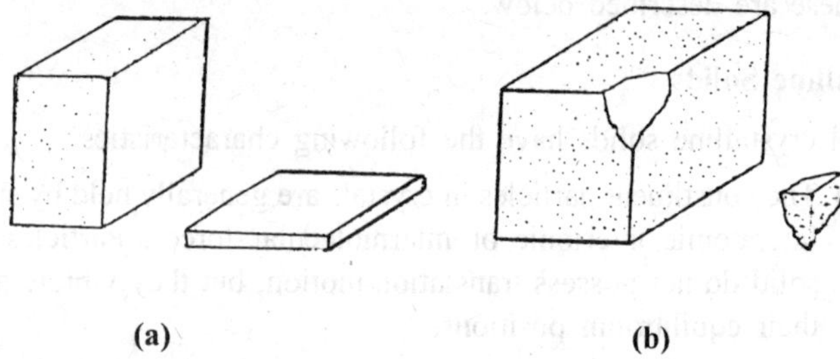

Fig. 2.2 : (a) Crystalline solids when hammered gently show a smooth cleavage along certain preferred directions. (b) The amorphous solids when hammered, show an irregular fracture in the direction of the force applied.

Polycrystalline Solids

In certain crystalline solids, the crystals are very find. These cannot be seen by the naked eyes. Such solids given an impression of being amorphous. Such fine crystalline solids, which appear as amorphous are termed as polycrystalline solids. Metal powders are polycrystalline in nature.

In a polycrystalline sample, there are small crystals. These crystals are randomly oriented. As a result, a sample of polycrystalline material appears to be isotropic, even when each individual crystal is anisotropic.

Amorphous Solids

Amorphous solids have some mechanical properties which are commonly associated with the word solid. But, the amorphous solids differ from the crystalline solids in many respects. So, amorphous solids are sometimes called as pseudo solids.

Some characteristic properties of the amorphous solids are;

(i) The mechanical, electrical and optical properties of amorphous solids do not depend upon the direction, *i.e.*, amorphous solids are isotropic. In this respect, the amorphous solids resemble liquids. It is for this reason that the amorphous solids are considered as supercooled liquids. There is only a short-range order in amorphous solids.

(ii) Amorphous solids do not poses sharp melting points. This indicates the absence of long-range order Amorphous soften on heating and gradually begin to flow like liquids.

(iii) Amorphous solids do not occur in characteristic geometrical shapes.

(iv) When hammered, amorphous material break in an irregular manner. When you hammer a glass sheet or a plastic sheet, these get broken in a very irregular way along many directions.

Examples. Some typical amorphous materials are Plastics, Glass, Rubber, Lamp black, pitch etc.

The solid state is next to gaseous state in its simplicity to understand, although it is also not free from complex problems. Solids are obtained by cooling liquids. There are inherent two possible ways of passing from the equilibrium liquid state into solid state namely crystallization and glass transition.

Crystallization is a transition from a state of short range order to one of long range order in the process of formation of a new phase. The concept of short range order and long range order is defined by the ratio between the distance over which the order extends and the dimensions of the elements arranged in the order. This is brought about by slowly cooling the liquid where the molecules, due to enough thermal energy and low viscosity of the medium, get enough opportunity to arrange themselves in a geometric order. With the slow release of energy, they grow on the nucleus maintaining the shape of the crystal. It is a phase transition of the first order.

Glass transition does not involve a change of phase, retaining a short range order. It is actually a super-cooled liquid in a non-equilibrium state, which solidify on cooling passing into glassy state. It loses the property of liquid state and acquires the properties of solid state. These changes do not occur abruptly but gradually over a certain range of temperature. The effect of heating an amorphous and crystalline solid is shown in Fig. 2.3. It can be seen that crystals have a sharp melting point while amorphous solids soften over a range of temperature.

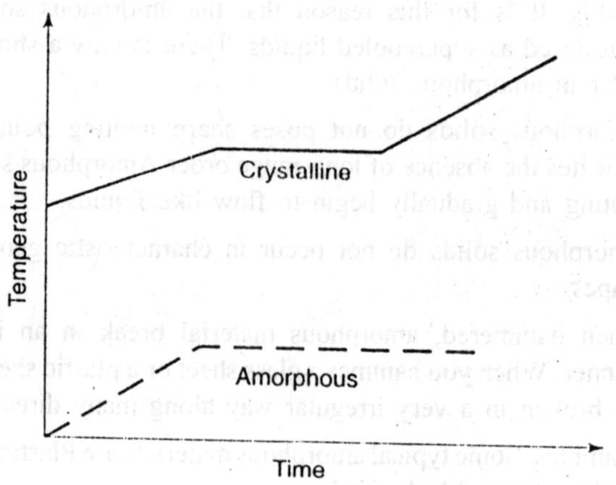

Fig. 2.3

The amorphous materials posses properties like mechanical, electrical, refractive index, etc., the same in all the directions, such materials are known as isotropic materials. The properties in a crystal depend on the direction along which they are measured. This shows an ordered arrangement of particles in a crystal, such substances are known as anisotropic materials.

A perfect crystal has a perfectly periodic structure which depends on the concentration of the solution and the rate of cooling. In actual practice, a perfect crystal is a model and not a physical reality.

TYPES OF FORCES

The forces that hold the atoms or molecules can be divided into three main groups:

1. Primary forces;
2. Secondary forces; and
3. Hydrogen bonding.

Primary Forces

As already discussed, it is the outermost valence electrons that play a part in bonding of atoms or molecules. These forces can be further sub-divided into (a) ionic (b) covalent (c) metallic, and (d) Van der Waals.

The above classification is very useful but in practice few bonds approach these types of bonds. By and large the bonds are of intermediate type as shown by Ketelaar in Fig. 2.4. The triangular diagram illustrates the transition between ionic, covalent, and metallic bonding.

Ionic Forces

If a sodium atom is placed near a chlorine atom the electron transference takes place resulting in a group Na^+ Cl^-. If a second positive sodium ion comes near a chlorine ion, it will experience a strong attraction towards Cl^- and weak repulsion from Na^+ as the force of repulsion diminishes rapidly with increasing distance as can be seen from potential energy curve (Fig. 2.5a) so it would be possible to add another Na^+ around the Cl^-. But the sodium ions must not come too close to each other and at the same time they should not try to contact the Cl^-. This repulsion interaction between the sodium ions and stearic hinderance will not allow more than a certain number of positive ions (6 in the case of sodium) around a negative ion. Once this group of seven ions is completed, the grouping of negative ions around the external positive ions will start to restore the electrical neutrality of the aggregate.

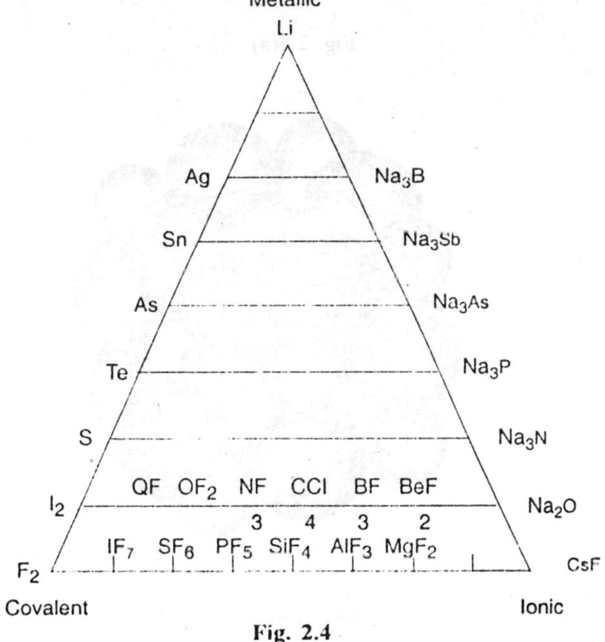

Fig. 2.4

The coulombic force between Na^+ and Cl^- is a short-range force. The homopolar bond works only in a particular direction while the ionic bond works in all the direction resulting into a space model, each ion surrounded by oppositely charged ions (Fig. 2.5b).

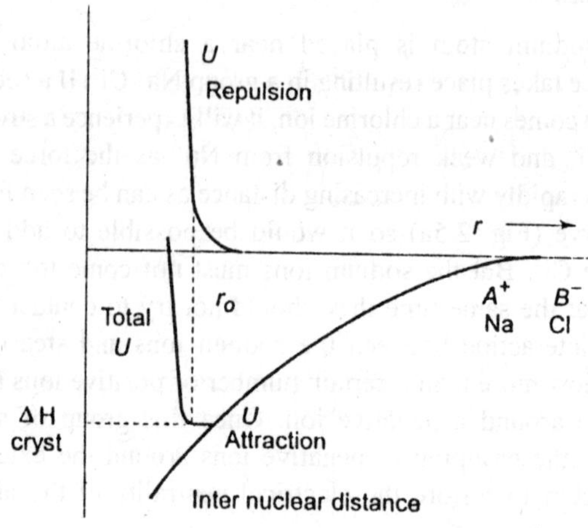

Fig. 2.5(a)

Fig. 2.5(b)

Covalent Forces

The most characteristic property in covalent bonding is the directional nature of the forces. Crystals of diamond, silicon, germanium, grey tin belong to this type of bonding. These are of two types—polar and nonpolar. The compounds containing –OH, –COOH, –Cl, –F groups are known as polar compounds like water, acetic acid, alcohol, etc. Among the most polar polymers are poly (vinyl alcohol), cellulose, starch, poly (acrylic acid), poly (methacrylic acid), poly (acrylonitrite), poly (vinylchloride), etc. The nonpolar molecules are CCl_4, benzene and polymers are polyethylene, polypropylene, polybutadiene, polyisoprene and polyisobutylene. Most of these polymers used to give amorphous structures, but with the advent of new catalysts by Nata, it has been now possible to produce the non-polar polymers as purely crystalline compounds.

Van der Waals Forces

In non-polar molecules, it is only the Van der Waals attraction that comes into play. The Van der Waals attraction is reduced to quantum mechanical effects due to the fluctuation of electrons in atoms. The fluctuation of electrons causes forming of rapidly fluctuating dipoles resulting in mutual attraction. These forces are non-directive and the binding energy is proportional to $\frac{\alpha^2}{d^6}$ where d is the internuclear separation and α the polarizability. The effect caused by these forces also depends upon the geometry of the molecules involved, for example, the flexibility of the polystyrene chain is due to arrangement of benzene ring in stacked position rather than side by side. These forces are also responsible for the interaction between chains both organic and inorganic which are normally at a distance of about 3-4 Å, *e.g.*, graphite, polyethylene. Van der Waals contribution to the total bonding energy in the alkali halides is small (5 Kcal/mole) because of the high symmetry of coordination. These forces are very important in layer structures like graphite, talc, asbestos. Although the configuration of the above compounds should be unstable due to coluombic repulsion but Van der Waals contribution stabilizes them. In case of $PdCl_2$ we observe a chain structure (Fig. 2.6).

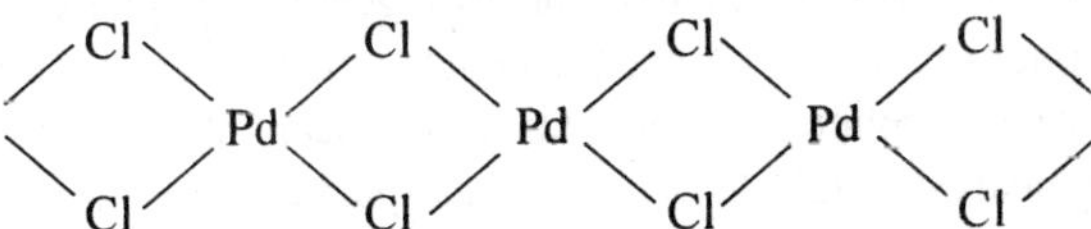

Fig. 2.6 : Planar chains in crystalline $PdCl_2$.

Ionic Solid

In ionic solids positive and negative ions occupy the lattice points. In sodium chloride some of the lattice points are occupied by Na and the others by chloride ions. It has been observed that molten sodium chloride conducts electricity. It is reasonable to assume that the ions which produce this conductivity actually exists as charged ion in the lattice. Further the energy required to separate the ions in the crystal lattice to an infinite distance agrees with that calculated on the assumption that the units are ions held together, by electrostatic forces. So the melting point of ionic crystals is high. (NaCl 800°, KCl 790°). Since there is no fixed directed force of attraction, they are brittle and have very little elasticity and can not be easily bent or worked. They have a great tendency to fracture by cleavage. The ions are not free to move in a solid state, hence, bad conductors of electricity but when melted becomes good conductors of electricity. In ionic crystals, some of the atoms may be held together by covalent bonds to form ions occupying definite positions and orientations in the crystal lattice, *e.g.*, in calcium carbonate a carbonate ion does not belong to a given calcium ion but three particular oxygen atoms do belong to a given carbon atom.

CLOSEST PACKED STRUCTURE

There are many solid elements and compounds whose atomic arrangements can be visualized in terms of the closest packing of identical atoms or ions so as to use the available space most efficiently and attain maximum density.

Packing Factor

The number of atoms per unit volume is known as packing factor.

Coordination Number

The coordination number is the number of closest particles in the neighbourhood of particle in question.

Spheres in Closest Packings

A closest packing of like spheres is an arrangement of spheres which permits each sphere to be in intimate contact with the largest number of its neighbours. For spheres of equal size, this means that each sphere would have six nearest neighbours in a closest-packed layer and twelve in a three-dimensional closest packing. In Fig. 2.7(a, b), we have shown a layer of spheres arranged in on the points of a square lattice. The efficiency of this arrangement where the area belonging to each circle is a square is calculated as follows:

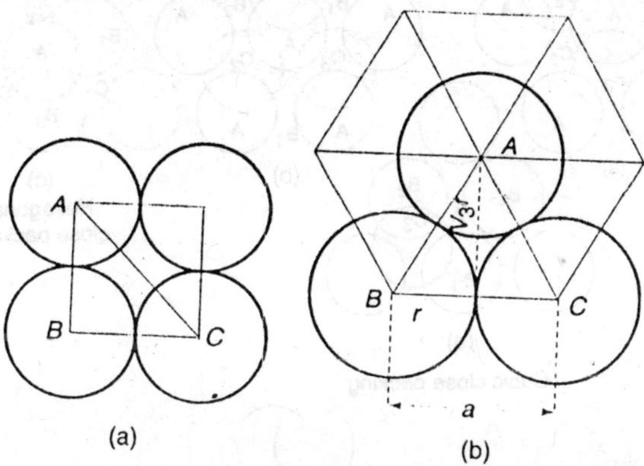

Fig. 2.7

$$\text{Efficiency} = 100 \times \frac{\text{Area of circle}}{\text{Area of a square}} = \frac{100\pi R^2}{4R^2} = \frac{100\,\pi}{4} \qquad ...(1)$$

Efficiency = 78.5 per cent.

The area associated with a circle here is a hexagon and each hexagon comprises of 12 right-angled triangles is $(R/2)\,(R/\sqrt{3}) = \frac{R^2}{2\sqrt{3}}$. The area of a hexagon is, therefore,

$$12 \times \frac{R^2}{2\sqrt{3}} = 2\sqrt{3}\ R^2. \qquad ...(2)$$

$$\text{Efficiency} = 100 \times \frac{\text{Area of circle}}{\text{Area of a square}} = \frac{100\pi R^2}{2\sqrt{3}\ R^2} = 90.7 \text{ per cent.} \qquad ...(3)$$

The packing arrangement in Fig. 2.7(c,d) represents the most efficient way of occupying space and this layer is called a hexagonal closest-packed layer.

In the two-dimensional hexagonal closest-packed arrangement of spheres, each sphere is surrounded by six others (nearest neighbours). Let us label each occupied site in the layer with letter A as shown in Fig. 2.8(a). The second layer of closest-packed spheres can be added by placing spheres in the depressions or voids (or holes) of the first layer. We shall label these voids by letters B and C (Fig. 2.8(a)). If we place

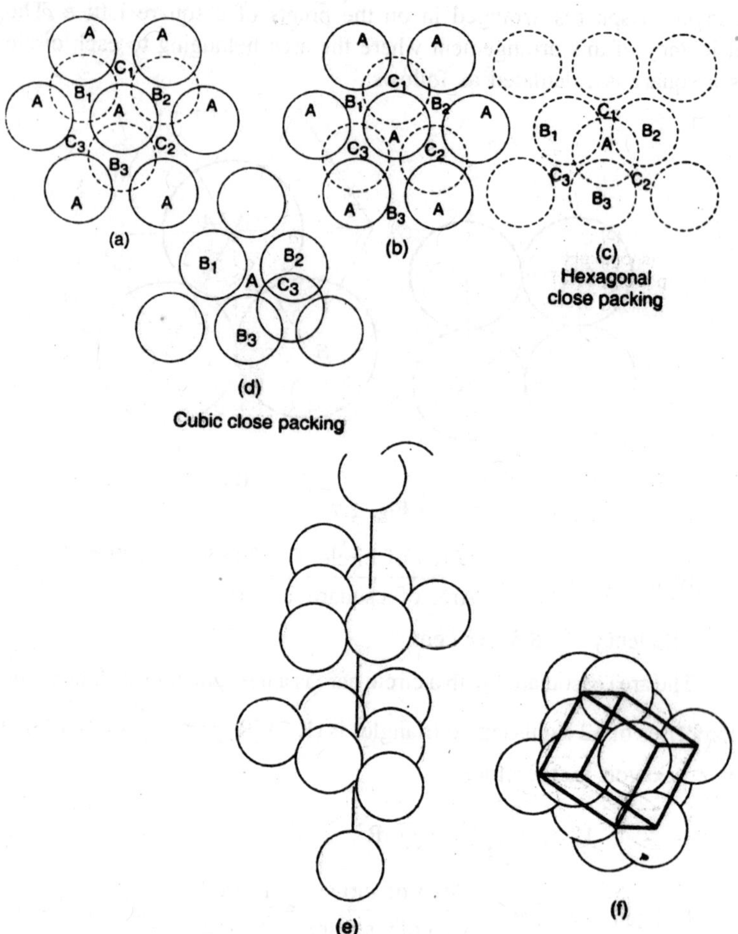

Fig. 2.8

a sphere at voids labelled B we cannot place any spheres at the C voids and vice versa. In Fig. 2.8(b) is shown the closest-packed arrangement obtained by packing spheres on B voids. The third layer can however, be added by placing spheres on the voids marked C or the A voids created by the second layer. If we place the spheres on C, ABC sequence is obtained, if we place them on A, AB sequence is obtained. These two sequences can be repeated indefinitely to get the ABCABCABC ... and ABABAB ... arrangements; these are respectively called the cubic closest packing (CCP) (Fig. 2.8d) and hexagonal closest packing (HCP) (Fig. 2.8c) arrangements respectively. The same arrangement is shown in Fig. 2.8e, f, g, h on enlarged scale. Both these are equally efficient with 74% of available volume occupied by spheres. In both CCP and HCP structures, coordination number of a sphere remains 12. In the CCP arrangement, there is always a sphere at each face centre of a cube and CCP is, therefore, also called a face-centred cubic arrangement. Metals are generally found to crystallize with either of these arrangements; in addition, metals also crystallize in the body centred cubic structure which is not a closest-packed arrangement. This arrangement has a coordination number of 8 and only 68% of the total volume is occupied by spheres.

Voids in Closest-Packed Structures

Two kinds of voids occur in closest packings. If a triangular void in a closest-packed layer has a sphere over it we get a void with four spheres around it and such a void is called a tetrahedral void.

In Fig. 2.8(a), the B voids (over which we place the second layer spheres) are tetrahedral voids. In Fig. 2.10(a) we have depicted a tetrahedral void. The coordination number of a tetrahedral void is four (Fig. 2.8b). If a triangular void pointing up in one closest-packed layer is covered by a triangular void pointing down in the next layer as shown in Fig. 2.11 (b), then the void is surrounded by six spheres arranged on

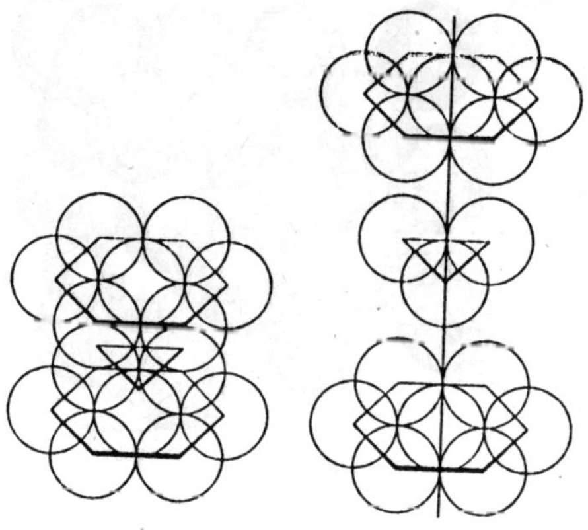

Fig. 2.9

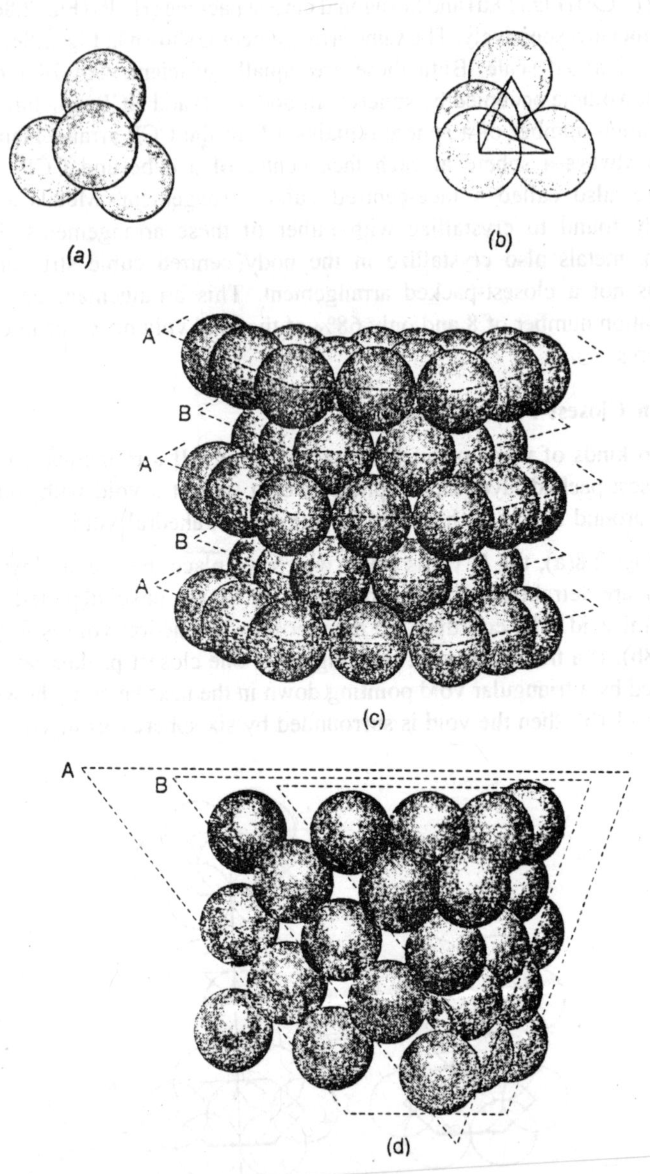

Fig. 2.10

the comers of an octahedron Fig. 2.11(b); such a void is called an octahedral void. The coordination number of an octahedral void is six as shown in Fig. 2.11(a). In Fig. 2.8(b), the C voids after the addition of the second layer represent octahedral voids. The same is shown in Figs 2.10 (c & d).

The number of octahedral voids belonging to one sphere is given by the ratio,

$$\frac{\text{Number of otahedral voids around a sphere}}{\text{Number of spheres around a void}} = \frac{6}{6} = 1 \qquad ...(4)$$

The number of tetrahedral voids belonging to a sphere is given by the ratio

$$\frac{\text{Number of tetrahedral voids around a sphere}}{\text{Number of spheres around a void}} = \frac{8}{4} = 2 \qquad ...(5)$$

We see that in a closest-packed arrangement, the number of octahedral voids is equal to the number of spheres while the number of tetrahedral voids is double the number of spheres.

Structures of Substances Related to Closest-Packed Lattices

Besides noble gases and metals which crystallize in closest-packed structures, a large number of inorganic compounds have structures related to closest packings. In such compounds one kind of atom forms the closest-packed structure while another kind of atom goes into the voids. In many of the solid solution of metals, similarly, some of the atoms are present in the voids. Properties of materials are often affected when foreign impurities or some of the atoms comprising the material itself are present as interstitial atoms in the voids; such interstitials are an important kind of imperfection in solids.

From the description of the closest-packed structures, we draw some inferences regarding the structures of simple compounds. For example, if in a compound of the general formula AB, the atom (or ions) form a close-packed lattice, then A atoms (or ions of opposite charge) can occupy all the octahedral voids or one-half of the tetrahedral voids. If the formula of the compound is A_2B, all the tetrahedral sites will be occupied by A atoms. Typical descriptions of ionic solids forming nearly close-packed structures are given below:

It must be remembered that the above descriptions are not exact in the sense that some of these ionic compounds are not strictly the closest-packed structures. For example, in NaCl the ions along the diagonal of the cube face do not touch each other.

An interesting case of packing is given by the oxide of iron, FeO, Fe_2O_3 and Fe_3O_4. FeO has the rock-salt (NaCI) structure and ideally Fe^{+2} ions should occupy the octahedral voids. However, this oxide is always non-stoichiometric and the composition is generally $Fe_{0.95}O$. We can obtain this composition if a small number of Fe^{+2} ions are replaced by Fe^{+3} ions in the octahedral voids. If we convert two-thirds of the Fe^{+2} ions into Fe^{+3} ions, we get the composition, $FeO.Fe_2O_3$ or Fe_3O_4; in Fe_3O_4, all Fe^{+2} ions are in octahedral voids and the Fe^{+3} ions are equally distributed between octahedral and tetrahedral voids. This is the well-known inverse spiral structure; another example of this structure is $MgFe_2O_4$ where Mg^{+2} replace Fe^{+2} of Fe_3O_4. In the normal spiral structure compounds of the general formula AB_2O_4 (*e.g.*, ferrites like $ZnFe_2O_4$; $MgAl_2O_4$), the divalent A^{+2} ions are in tetrahedral voids and the trivalent B^{+3} ions are in octahedral voids.

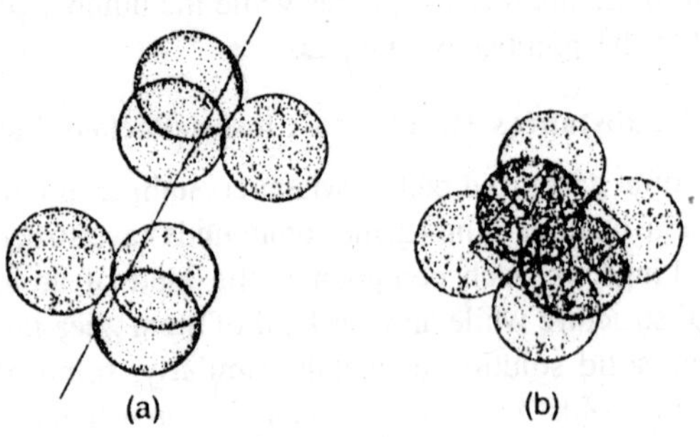

Fig. 2.11

We see from the earlier discussion that ionic sizes determine the nature of packing in the close-packed structure. Thus, in ZnS the small Zn^{+2} ions occupy the voids while the large S^{-2} ions form the close-packed lattice. In NaCI, on the other hand (where the ionic sizes are similar), Na^{+} ions occupy the octahedral voids. The coordination number of an ion (or an atom) in the close-packed structures depends on ratio of its radius with respect to the radius of the atoms surrounding it. In the case of ionic solids, this radius ratio is the ratio of the cation radius to the anion radius. Generally, in ionic solids, cations tend to get surrounded by the largest possible number of anions and therefore, larger the radius ratio larger is the coordination number of the cation. We have summarized below the range of radius ratios over which coordination numbers of 2, 3, 4, 6, 8 and 12 are likely to be stable. By employing value of ionic radii for cations and anions, it is possible to predict the coordination number of cations in compounds. Radius ratio is indeed a useful concept for prediction of structures of ionic compounds.

$$\text{Cubic (8)} \underset{}{\overset{0.732}{\rightleftharpoons}} \text{octahedral (6)} \overset{0.414}{\rightleftharpoons} \text{tetrahedral} \overset{0.225}{\rightleftharpoons} \text{triangularic} \overset{0.155}{\rightleftharpoons} \text{linear}$$

Metallic Bonding

Nearly eighty elements exists as metals at NTP. Metals differ from other substances in terms of their electronic configuration. Metals interact by sharing their electrons not between its two atoms but amongst all the atoms. Thus, a metal may be viewed as a collection of positive ions embedded in a sea of mobile electrons. Such a cloud of electrons often referred to as electron gas is responsible for most of their electrical properties.

Other Types of Bonding

The classification of solids into four major types should not be taken in simply that there is a sharp demarcation between one type and another. In some solids, more than one type of bonding are responsible for the stability of structures. As mentioned earlier. Van der Waals forces are responsible for inter-layer binding in graphite. Similarly, there are susbtances which form thread-like structures (*e.g.*, amorphous sulphur, asbestos) where atoms are linked to one another along a chain by covalent bonds, but the chains are held together by weak Van der Waals forces. In such substances, threads (or fibres) can be pulled out easily, but it is relatively harder to break threads. There are solids like the metal silicates where the bonding is between ionic and covalent bonds. Silicates form thread-type, sheet type or three-dimensional structures, held together by ionic interaction with cations present in them.

There are solids like ice which are held together by weak hydrogen bridges or hydrogen bonds formed between molecules (H_2O molecules in the case of ice). In proteins, stable chains of atoms are formed by covalent as well as hydrogen bonds but in the solid state the chains are held together by Van der Waals forces between them.

IONIC CRYSTALS

Solids in which ions are arranged in some definite geometrical pattern are called ionic crystals. These ions are bound by plane surfaces called faces. These faces intersect at an angle giving definite shape to the crystal. Steno observed that the shape of a crystal depends on the details of crystalization but at a given temperature the angle between planes of a crystal always remain constant. This is known as Stenos law. That is why a crystal on cleavage maintains its geometry. Sometimes a substance crystallizes out in different shapes. This is known as polymorphism.

Laws of Rotational Indices

The law of rotational indices states that the intercepts of the plane of various faces of crystal on a suitable set of axes can be expressed by small integral multiples of three unit distances.

Space Lattices

The diffraction of X-rays by crystals is a phenomenon of great importance. Diffraction studies provide much of the structural information on the arrangement of atoms, ions or molecules in crystals.

Hauy suggested that the environment about any particle in space, *i.e.*, in three direction is the same as that about any other particle. Thus, a regular three-dimensional array of points, representing particles, is known as space lattice.

Unit Cell

A unit cell in the crystal is the smallest repeating unit in three dimensions, which reproduces the whole crystal by the displacement of the unit along the direction of the three axes. The crystal may be considered to consist of a very large number of unit cells, each one in direct contact with its nearest neighbours and all are similarly oriented in space. The shape and size of a unit cell are given by the spacings along the three axes (a, b, c) and the three angles (α, β and γ) between the faces.

Zero Point Energy

A crystal is found to have some vibrational and translational energy at absolute zero. This is known as the vibrational energy at absolute zero or the zero point energy of the crystal.

Indices

The intercepts on the axes of X, Y, Z of a unit cell are a, b, c, and are known as Weiss indices. The reciprocal intercept $\frac{a}{oA}$; $\frac{b}{oB}$ and $\frac{c}{oC}$, *i.e.*, h, k, I are known as Miller indices (Fig. 8.10 a, b). They are always integral.

Weiss indices	2a,	b,	l/2c,	a,	b,	c,	a,	1,	1
Miller indices	1,	2,	4,	1,	1,	0,	1,	0,	0

Depending upon the magnitude of Weiss indices and the angles between them, there are seven different types of crystal system.

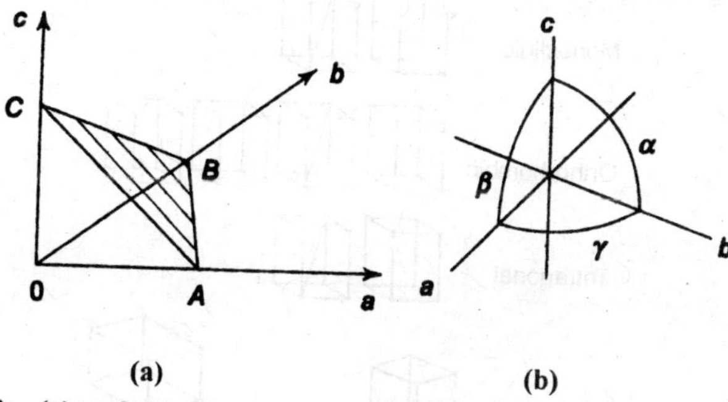

(a) (b)

Fig. 2.12 : (a) and (b) The axial cross illustrating the coordinates used in specifying unit cell and lattice dimensions

Bravais Lattices

In 1848, Bravais showed that there are fourteen ways of arranging points space so that each point has identical surroundings. These lattices are known as Bravais lattices.

Symmetry Properties

The arrangement of crystal faces such that the crystal transform into image of itself is known as symmetry operation. These operations are as a result of some symmetry elements. These symmetry elements are international symbols.

An axis of rotatory inversion is the combination of a rotation axis with a centre of symmetry. It is denoted by placing a bar over the symbol. The various symmetry elements are given in Fig. 2.14.

There are 32 possible combinations of these symmetry elements. These define 32 crystallographic point groups which determine the 32 crystal classes. The symmetry operations are reversible, *i.e.*, A(BC) = (AB)C. Only holohedral class of crystals follow the complete symmetry. Crystalline rocksalt and iron pyrites are cubic crystal. Rocksalt crystal possess the full symmetry of the cube but iron pyrites does not show all the symmetries because all the faces in iron pyrites are not equivalent.

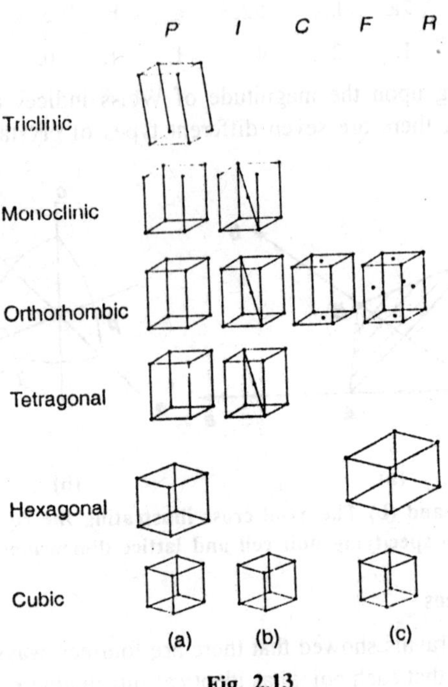

Fig. 2.13

Space Groups

The classification of crystals is based on the various groups of symmetry elements operations, such that they leave atleast one point in the crystal as invariant. Hence, they are called point groups. The possible three dimensional lattices symmetry operations of infinite figures are called space groups. It has been worked out that the internal symmetry of crystal may be classified according to 230 space groups, *i.e.*, there are only 230 ways of repeating a particular pattern.

Table 2.1

No symmetry	1	Four-fold rotation axis	4
Mirror plane of symmetry	m	Four-fold rotatory inverter	4
Two-fold rotation axis	2	Six-fold rotation axis	6
Three-fold rotation axis	3	Six-fold rotatory inverter	6
Three-fold rotatory inverter	3	Centre of symmetry inverter	1

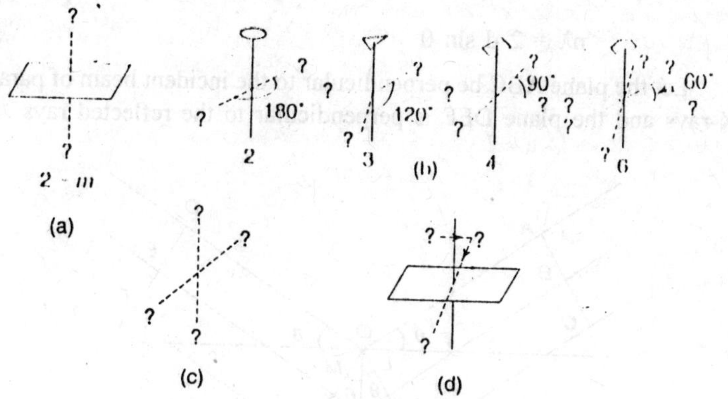

Fig. 2.14

Diffraction Methods

In 1912 Von Lave suggested that a crystal can act as a three-dimensional diffraction grating for X-rays of wave length ranging 0.001-50Å. X-rays are produced by striking atoms in a solid target by accelerated electrons. Thus, the electrons in the inner orbit of the target are displaced by bombardment and the energy liberated by the transition of electrons from a higher energy level to the vacancy in the inner orbit is given out in the form of an x-ray photon. These transitions produce characteristic lines in the x-ray spectrum. A copper target gives an intense line at 1.54Å. In addition to the above radiation, a continuous range of X-ray wave lengths are obtained. These are called white X-rays. Thus, using a nickel filter radiations of wave length 1.54Å can be picked up which are widely used for polymers.

Beams of electrons and neutrons has also been diffracted by crystals. Neutron diffraction determines the position of protons in a structure. X-rays scattered by electrons do not generally lead to a precise position of protons. Electrons do not penetrate very far into solids, but their use in gases has been found very valuable in finding the structure of gas molecules. Reflections from the surface reveal facts about their structure.

Hugens-Fresnel suggested that every point reached by the electromagnetic vibrations may be regarded as a radiation centre. Bragg suggested that it would be convenient to assume that the X-rays are reflected from the planes in the crystal. Woulfe-Bragg found a simple relation between A the wave length of the X-rays, d the distance between planes in a crystal and Q the angle of reflection

$$n\lambda = 2\ d \sin \theta \qquad ...(6)$$

Let the plane ABC be perpendicular to the incident beam of parallel X-rays and the plane DEF is perpendicular to the reflected rays

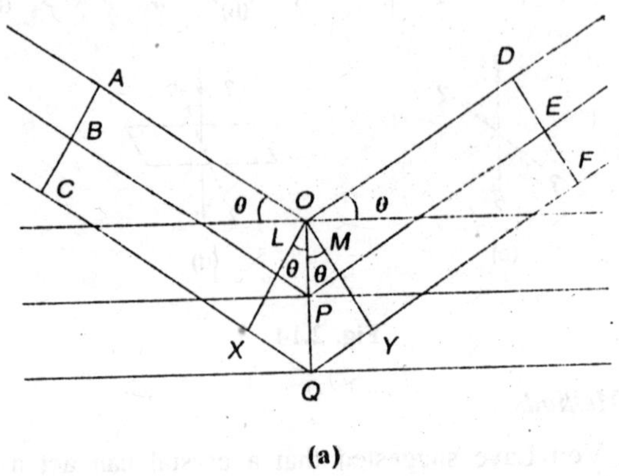

(a)

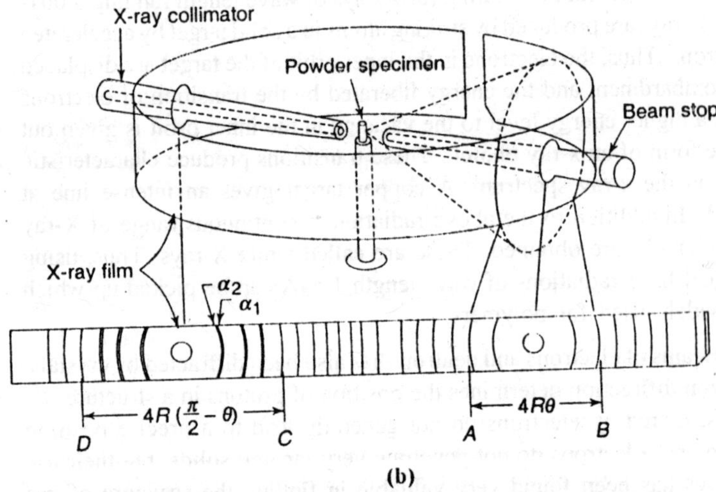

(b)

Fig. 2.15

It can be seen that the ray BPE travels farther from AOD by a distance equal to LP + PM. The angle of reflection is equal to the angle of incidence θ and LP & PM is equal to the integral number of wave lengths.

$$\therefore \quad LP + PM = n\lambda \qquad ...(7)$$

$$LP = PM = d \sin \theta \qquad ...(8)$$

$$\therefore \quad n\lambda = 2d \sin \theta \qquad ...(9)$$

When n = 1, it is called the first order reflection and n = 2, it is called the second order reflection, etc. Equation 8.4 can be written as

$$\lambda = 2\left(\frac{d}{n}\right) \sin \theta \qquad ...(10)$$

$$= 2\, d_{hkl} \sin \theta$$

where d is the distance between the planes having hkl as miller indices.

The second order reflections from 100, 110, 111 planes are labelled as 200, 220, 222 respectively. In a cubic crystal distance between planes with Miller indices hk*l* is given by

$$d_{khl} = \frac{a}{(h^2 + k^2 + l^2)^{1/2}}$$

when a is the length of the side of the unit cell.

Thus, the spacings between the planes in a primitive cubic crystals may have values a, $\frac{a}{\sqrt{2}}, \frac{a}{\sqrt{3}}, \frac{a}{\sqrt{2}}, \frac{a}{\sqrt{5}}, \frac{a}{\sqrt{6}}, \frac{a}{\sqrt{8}}$ but not $\frac{a}{\sqrt{7}}$ as $h^2\ k^2 + l^2$ cannot have a value.

In the case of face centred cubic lattice, the spacings which will appear are $\frac{a}{\sqrt{3}}, \frac{a}{\sqrt{4}}, \frac{a}{\sqrt{8}}, \frac{a}{\sqrt{11}}, \frac{a}{\sqrt{12}}, \frac{a}{\sqrt{16}}, \frac{a}{\sqrt{19}}, \frac{a}{\sqrt{20}}$, etc.

In the case of body centred cubic lattice the inter planer spacings are $\frac{a}{\sqrt{2}}, \frac{a}{\sqrt{4}}, \frac{a}{\sqrt{6}}, \frac{a}{\sqrt{8}}$, etc.

Thus, knowing the missing X-ray lines it would be possible to predict the type of cubic lattice of a given crystal.

If X is known and the direction of θ in which the intensity of the scattered radiations is maximum is determined experimentally, d can be calculated and a can be known from the density of the cubic crystal.

Example:

The length of the side of a unit cell of potassium is 5.34Å. Calculate the distance between 200, 110 and 222 planes.

$$d = \frac{5.34}{\left(h^2 + k^2 + l^2\right)^{1/2}}$$

For 222 planes d is $\frac{5.34}{(4)^{1/2}} = 2.67Å$

For 110 planes d is $\frac{5.34}{(2)^{1/2}} = 3.77Å$

For 222 planes d is $\frac{5.34}{(12)^{1/2}} = 1.54Å$

Macromolecules : In macromolecules, there are three basic types of ordering; the small period, the helicals when repeat distance is along the chain and the large period.

The small period characterizes the dimensions of the smallest structural elements out of which all polycrystalline bodies are made in their various combinations.

Regularly built macromolecular chains with side substitutes assume in space helical conformation with constant pitch. The helical pitch conforms to the periodicity along the chain with a specific size of the order of several angstroms.

A special type of ordering inherent only to polymer bodies is characterized by the alternate regions of long-range and short-range orders and in the limit; amorphous and crystalline regions are obtained in oriented polymers. The large periods have the order of hundred of angstroms. Three methods are used with crystals.

1. Laue's Method

White x-rays are used in this method. The x-ray pattern obtained represents a system of spots from various systems of planes in a crystal. This is not used for polymers because polymer single crystals of large size are usually not obtained.

2. Bragg's Method

In this method a monochromatic radiation is used and the sample is rotated or tilted when it is photographed. The photographic film is arranged on the surface of the cylinder around a sample which is placed along the cylinder axis. Diffraction maxima appear in the coarse of rotation and then θ can be measured.

3. Debye Scherrer Method

Samples in the form of powder or a polycrystalline body are examined. The incident monochromatic X-ray is diffracted on the planes of a crystallite of a polycrystalline sample that satisfy the Woulfe-Bragg equation by their orientation with respect to incident ray. Diffracted rays will spread along a cone with the vertex angle equal to 2θ. The intersection of these cones with the flat photographic film placed perpendicular to the incident ray behind the sample produces a system of concentric rings of growing radii, each of which contains all the reflections with the same angle θ. The reflections are recorded on circular photographic film. If coarse crystals are used, the powder pattern produces rings of spots, each spot produced by the different crystal planes. If the crystals are very fine, continuous arcs are obtained on the film (Fig. 2.15b).

X-ray diffraction analysis is used in obtaining the structure of polymeric bodies and its change as a result of heat, mechanical and other effects, on the phase transformation of a body and to determine the conformation of chains. They also give average dimensions of crystal and their distribution according to dimensions. They make it possible to detect the origination of the smallest (10-100 A) cracks in polymeric bodies.

Primitive Cubic Lattice

In a primitive or simple cubic cell, the cations are located only at the corners of the unit cell and its centres occupy the anions, *e.g.*, Cs Cl, *i.e.*, it contains one ion pair per unit cell. Each cation is therefore, surrounded by eight anions and vice versa (Fig. 2.16a & b). This arrangement is known as cubic coordination. This type of structure is exhibited by cesium and thalium chloride, bromide and iodides and by a host of other compounds. The packing factor is 0.52.

$$\text{P.F.} = \frac{\frac{4}{3}\pi r^3}{a^3} = \frac{4\pi}{3}\frac{(r)^3}{8r^3} = 0.52 \qquad ...(11)$$

Face Centred Cubic Lattice

This is known as rock salt structure. There are four ion pairs or molecules in a unit cell of sodium chloride. The ions at the corners are shared by eight unit cells while those on the edges by four unit cells (Fig. 2.16a, b).

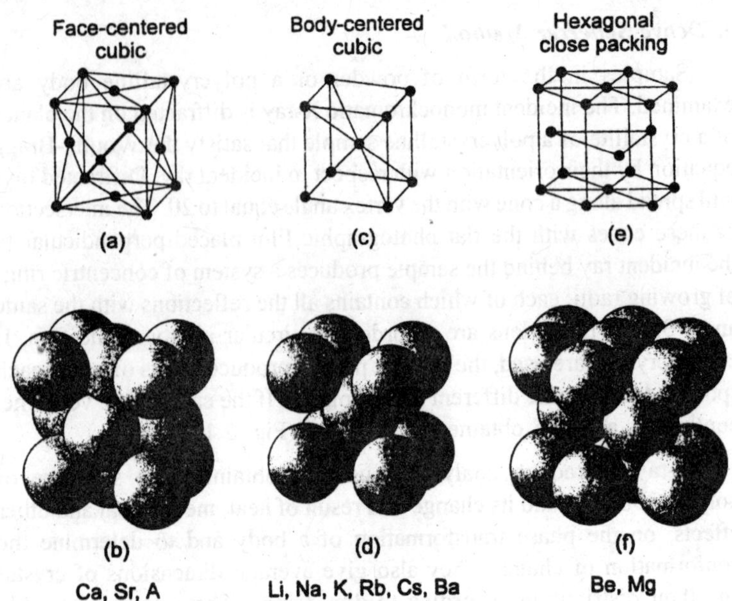

Fig. 2.16

Considering the pattern made by cations, it is found that (Fig. 2.16a) they occupy all the corners of a cell and centres of the faces of the unit cell. Therefore, it is called a face centred unit cell. Similarly, the anions will also have a face centred unit cell. It has a octahedral coordination with a total coordination number of 12, 6 for the cation and 6 for the anion. So each ion is surrounded by six other ions (Fig. 2.16g & h).

This is a very common structure and is assumed by alkali halides, most of oxides, sulphides, selenides and tellurides of the alkaline earth metals. Packing factor (P.F.) is given

$$\text{P.F.} = \frac{4}{3}\pi r^3 \times \frac{4}{a^3} = \frac{\pi r^3}{\left(2\sqrt{2}\right)^3 r^3} = 0.74 \qquad ...(12)$$

Body Centred Cubic Lattice

Most of the metals show a body centred crystalline lattice Fig. 2.16c & d. The alkali metals, barium, iron, chromium, molybdenum are some of the examples. There are eight atoms at the corners of the unit cell and one is at the centre. Hence, the coordination number is eight. Goldschmidt derived an emperical ratio between the metallic radius and the coordination number.

Coordination number	12	8	6
Atomic radius ratio	1.00	0.97	0.96

$$P.F. = \frac{4}{3}\pi r^3 \times 2a^3$$

$$= \frac{8}{3}\pi r^3 \times \frac{3\sqrt{3}}{64r^3} = 0.68 \quad ...(13)$$

The structure of titanium (and arsenic) is shown in Fig. 2.17a & b. It is composed of octahedra each with a titanium ion at the centre and the octahedra shares edges and corners with each other.

The structure of arsenic is shown in Fig. 2.17b. It is a NaCl type of cell with all the atoms of the same kind. It is deformed along a three-fold axis which gives a corrugated layer structure formed by pyramidal groups of atoms.

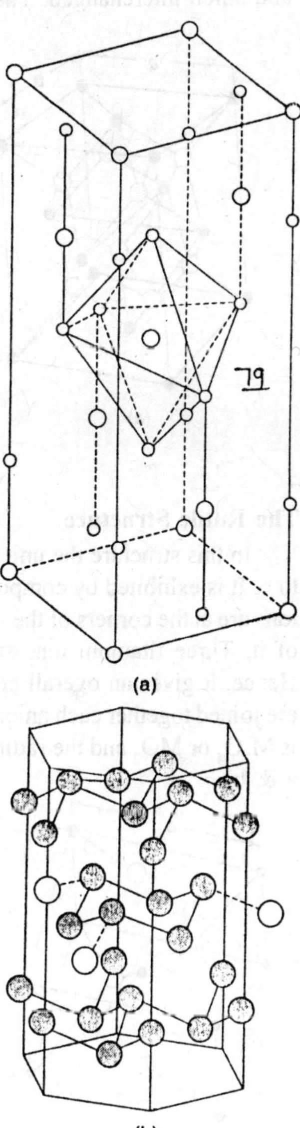

(a)

(b)

Fig. 2.17

Fluorite Structure

In the fluorite each cation is surrounded by eight anions and each anion is surrounded by four cations to give compounds of the type AX_2. The anions have a tetrahedral coordination with coordination number 4, while cations are surrounded by eight anions at the corner of a cube giving a cubic coordinations. The overall structure is known as 8:4 coordination. The compound will have a formula MX_2 (Fig. 2.18a & b).

Calcium fluoride, and difluorides of strontium and barium, etc. give fluorite structure. Their radius ratio lies between 0.73 to 1.00.

The oxides, sulphides, selenides, tellurides of lithium, sodium, potassium give antifluorite structure which is the same with positions of cations and anion interchanged. The formula of their compounds will be M_2X.

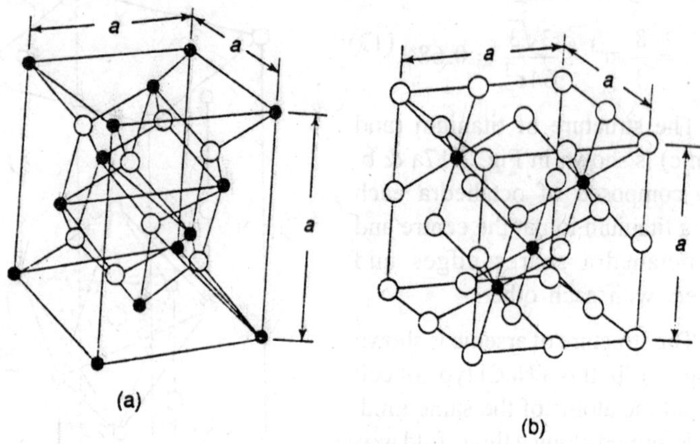

Fig. 2.18

The Rutile Structure

In this structure the unit cell is tetragonal where a = b but not equal to c. It is exhibited by compounds like titanium dioxide. The six oxygen ions are at the corners of the octahedron and the titanium ion at the centre of it. Three titanium ions surrounding each oxygen ion are coplaner. Hence, it gives an overall coordination of 6 : 3. When these octahedra are joined together each anion forms part of three octahedra. The formula is M_2O_4 or MO_2 and the radius ratio will be greater than 0.414 (Fig. 2.19 a & b).

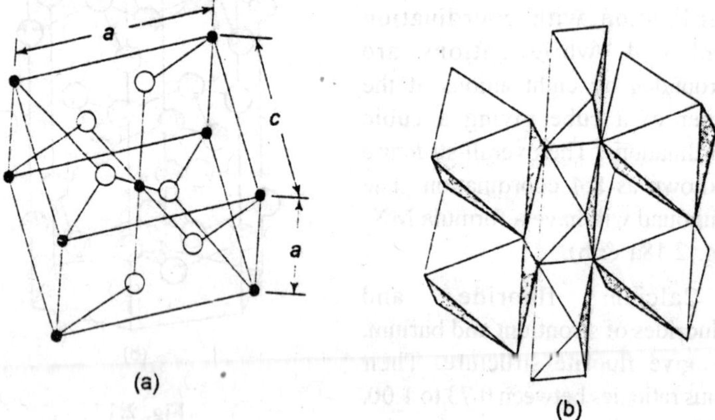

Fig. 2.19

Oxides of tin, lead, etc., and difluorides of transitional metals give rutile structure on crystallization.

Zinc-Blend Structure

Another important structure is zinc sulphide lattice. Zinc sulphide crystallizes in two structures, sphalerite lattice and wurtzite lattice. The wurtzite structure differs from sphalerite in being tetragonal rather than cubic. Sulphide ions form a face centred cubic lattice with zinc ions occupying alternate tetrahedral sites. Both zinc and sulphur have coordination number four and the radius ratio is less than 0.414 (Fig. 2.20a & b).

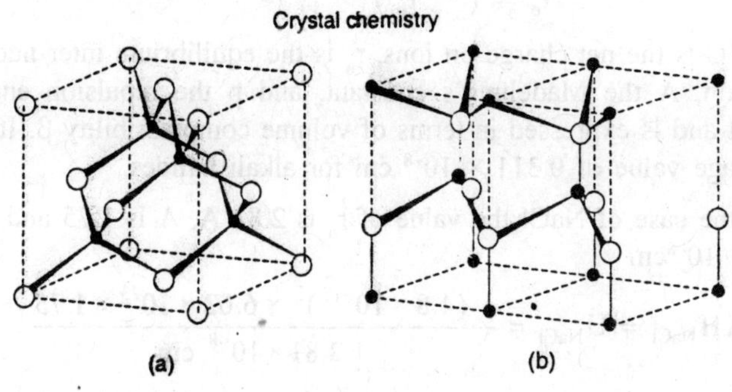

Fig. 2.20

Zinc sulphide, halides of copper, silver iodide, beryllium sulphide and calcium fluoride crystal have zinc blende structure. They form 1 : 1 compound in which cation is much smaller than the anion.

CRYSTAL STRUCTURE

Bragg during his study of X-ray diffraction derived a relation between the distance between planes in a crystal and the angle at which the reflected radiations have a maximum intensity.

$$\lambda = 2\left(\frac{d}{n}\right)\sin\theta = 2\, d_{hukl} \sin\theta \qquad ...(14)$$

where n is the order of reflection usually set equal to 1, d is the distance between planes, θ is the angle and λ is the wave length employed.

In a cybic system $$d_{hkl} = \frac{a}{\left(h^2 + k^2 + l^2\right)^{1/2}} \qquad ...(15)$$

$$= \frac{2a \sin\theta}{\left(h^2 + k^" + l^2\right)^{1/2}} \quad ...(16)$$

thus, knowing h, k, l, θ and d_{hkl}, a can be calculated.

The Cohesive Energy of Crystals

Born Mayer derived the equation for crystal energy per mole

$$U = \Delta H_{cryst} = \frac{Z{+}Z{-}C^2NA}{r_o}\left(1 - \frac{\rho}{r_o}\right) \quad ...(17)$$

$$= \frac{Z^2 \varepsilon^2 A}{r_o}\left(1 - \frac{\rho}{r_o}\right)$$

where ZC is the net charge on ions, r_o is the equilibrium inter nuclear separation, A the Madelung's constant, and p the repulsion energy constant and is expressed in terms of volume compressibility β. It has an average value of 0.311×10^{-8} cm for alkali halides.

In the case of NaCl the value of r_o is 2.81 A, A is 1.75 and ρ is 0.311×10^{-8}cm

$$\Delta H_{NaCl} = U_{NaCl} = -\frac{(4.8 \times 10^{-10})^2 \times 6.02 \times 10^{23} \times 1.75}{2.81 \times 10^{-8}\ \text{cm}}$$

$$\left(1 - \frac{0.311 \times 10^{-8}\ \text{cm}}{2.81 \times 10^{-8}\ \text{cm}}\right)$$

$$= 7.68 \times 10^{12}\ \text{crg/mole}$$

$$= -184\ \text{K cal/mole}$$

which is in good agreement with the experimental value of –183.3 Kcal/mole.

Compressibility

The compressibility β is given by Born and Landes equation

$$\frac{1}{\beta} = -\frac{mnu_o}{9\upsilon_o} = -\frac{nu_o}{18r_o^3} \quad ...(18)$$

$$\therefore \quad n = 1 + \frac{18r_o}{\beta AZ^2\varepsilon^2} \quad ...(19)$$

The lowest potential energy per molecule is $U_o = \frac{AZ^2\varepsilon^2}{r_o}\left(1 - \frac{1}{n}\right)$

where n is the repulsive force constant ε the charge, U_o is molecular volume equal to $2r_o^3$.

Born-Haber Cycle

Bom-Haber made the direct determination of the crystal energies. In the case of sodium chloride

$$Na^+(g) + Cl^-(g) \xrightarrow{\Delta H^o_{Cyst}(NaCl)} NaCl\ (cryst)$$

$$-\Delta H^o_{ion}(Na, g)\downarrow \qquad \downarrow\Delta H^o_{ion}(Cl^-, g)$$

$$\downarrow\Delta H^o_f(NaCl)$$

$$Na(g) + Cl(g) \xrightarrow[-\Delta H^o_f(Cl,g)-\Delta H^o_f(Na,g)]{} Na(s) + 1/2\ Cl_2(g)$$

$$\therefore \Delta U = \Delta H^o_{cryst}\ (NaCl) = \Delta^o_f H_f\ NaCl\ (cryst) - \Delta H^o_f Na(g) - \Delta H'_{ion}\ Na^+(g)$$

$$-\Delta H^o_f Cl(g) + \Delta H^o_{ion}\ Cl^-(g)$$

$$= (-98.23 - 25.9 - 119.9 - 29.01 + 86.8)\ Kcal$$

$$= -186.2 Kcal$$

Δ^o_f NaCl (cryst) = the crystal energy

Δ^o_f NaCl = standard heat of formation of NaCl (s)

Δ^o_f Na(g) = heat of sublimation of Na (s)

Δ^o_f Cl(g) = energy of dissociation of Cl_2 (g) into atoms

Δ^o_{ion} Na^+ (g) = Ionization potential of Na

Δ^o_{ion} Cl^-(g) = electron affinity of Cl

Structure of Simple Ionic Compounds

It is much more difficult to describe the crystal structures or unit cells of compounds compared to those of elements. In the case of an element like copper (FCC), all the lattice points are occupied by copper atoms. However, if we were to consider the structure of the simplest ionic compound of the formula AB, we will have to describe the arrangement of both A and B ions. We shall now briefly examine the structures of a few simple ionic compounds of the type AB and AB_2.

Compounds of the type AB are generally cubic and can possess one of the three types of structures: cesium chloride (CsCI) (Fig. 2.17a) type, rock salt (NaCI) type and zinc blende (ZnS) type.

Table 2.2

	Description and examples	*Coordination number*	*Number of molecules per unit cell*
CsCl-type	Cs^+ at the body centre of a cube of Cl^- and Cl^- at the centre of a cube of Cs^+, *e.g.*, CsCl, CsBr; Csl, TIC1.	Cs^+8 Cl^-8	1
Nad-type	Both Na^+ and Cl^- form FCC lattices and the two interpenetrate each halfway, *e.g.*, All sodium halides, AgCl, Transition metal monoxides.	Na^+ 6 Cl^- 6	4
ZnS-type	This also has a face centred arrangement of ions, but more complex than NaCI,S^{-2} 4 ZnS, CuCl.	Zn^{+2} 4	4

Chlorides, bromides and iodides of lithium, sodium, potassium and rubidium (as well as some of the halides of silver), all of which ordinarily possess the NaCl structure with 6:6 coordination, transform to the CsCl structure with 8 : 8 coordination on application of pressure; notice that high pressure increases the coordination. On the other hand, CsCI on heating transforms to the NaCl structure at 490°C; similarly, rH_4Cl, NH_4Br and NH_4I transform from the CsCl structure to the NaCI structure at 184°, 138° and –18°C respectively.

Many of the compounds of the formula AB_2 crystallize in the cubic fluorite (CaF_2) (Fig. 2.18) type structure or the tetragonal rutile (TiO_2) structure (Fig. 2.19). In the CaF_2 structure, each F" ion is coordinated by four Ca^{2+} ions and each Ca^{+2} ion by eight F^- ions (cubic coordination). Other examples of compounds with CaF_2 structure are SrF_2, CdF_2; PrO_2 and ThO_2. In the rutile structure, the titanium atoms form a body-centred tetragonal structure and each titanium is surrounded by three titanium ions 3 : 6 coordination). Typical examples of rutile phases are RuO_2, SnO_2, PbO_2 and MgF_2.

The structures of many of the ionic compounds can be understood in terms of the packing arrangements of ions (and the relative sizes of cations and anions).

Ionic Radii

In the absolute sense of the word it is not a constant parameter. It is not possible to find the exact ionic radii as ions are not rigid bodies, but for all practical purposes if we take half the distance of approach between two similar ions, as the ionic radii, it will give a very useful data to build up the concept of the ionic structure in the compounds. The sum of the ionic radii of both ions in compound gives idea of the approach distance between two ions, which in turn gives the idea of their geometrical arrangement. The radius of an ion remains the same in various crystals as the repulsive force increases very sharply as the internuclear distance becomes smaller than a certain value.

In Fig. 2.21a four anions are arranged around a cation. This is the limiting condition of stability as the anions are touching the cations.

$$XY^2 + YZ^2 = XZ^2 \quad ...(20)$$

$$(2r_a)^2 + (2r_a)^2 = (2r_a + 2r_c)^2$$

or $$r_a + r_c = \sqrt{2}\ r_a \quad ...(21)$$

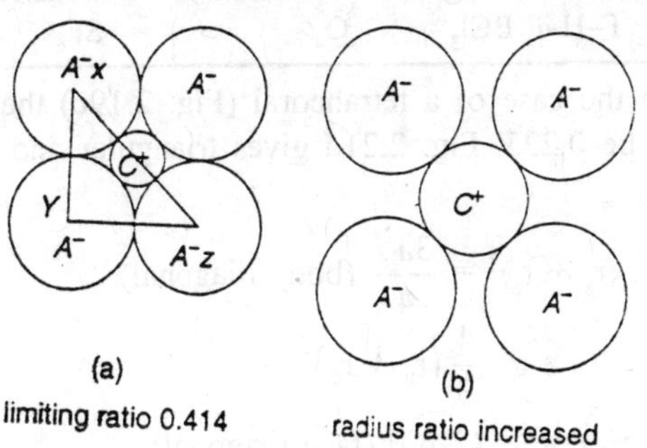

(a) limiting ratio 0.414

(b) radius ratio increased

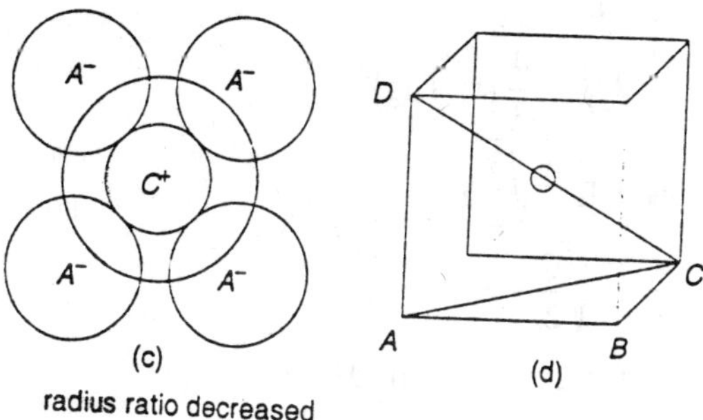

(c) radius ratio decreased

(d)

Fig. 2.21

or $r_c = r_a(\sqrt{2} - 1) = r_a . 414$

$$\frac{r_c}{r_a} = 0.414 \quad ...(22)$$

In a three-dimensional arrangement it will represent an octahedral arrangement with two additional ions one above and one below the cation. By increasing or decreasing the radius ratio various arrangements are obtained. If the ratio is decreased (Fig. 2.21c), the anions would tend to arrange tetrahedrally. If the ratio is increased above 0.73 a cubic arrangement is obtained (Fig. 2.21b), *e.g.*, CsCl.

Table 2.3

Radius ratio	0.15	0.15-0.23	0.23-0.41	0.41-0.73	0.73-1.0	71.0
Coordination No.	2	3	4	5	8	12
Shape	Linear	triangular	Tetrahedral		Octahedral	Cubic
Example	F–H–F	BCl_3	O_4		SF_6	CS Cl

Similarly, in the case of a tetrahedral (Fig. 2.19c) the maximum radius ratio will be 0.225. Fig. 2.21d gives triangular and tetrahedral arrangements

$$\text{In DOAC, } (r_a + r_c)^2 = \frac{3a^2}{4} \text{ (body diagonal)}$$

$$a^2 = \frac{4}{3}(r_a + r_c)^2$$

From ABC, $a^2 + a^2 = (2\,r_a)^2$ (face diagonal)

$$a^2 = 2r_a^2$$

$$2r_a^2 = \frac{4}{3}(r_a + r_c)^2$$

$$r_a + r_c = \sqrt{\frac{3}{2}}\,r_a$$

$$\frac{r_c}{r_a} = \sqrt{\frac{3}{2}} - 1$$

$$= 1.225 - 1 = 0.225.$$

Generally in ionic solids, cations tend to get surrounded by the largest possible number of anions and therefore, larger the radius ratio larger is the coordination number of the cation. By employing value of ionic radii for cations and anions, it is possible to predict the coordination number of cations in compounds. Radius ratio is indeed a useful concept for prediction of structures of ionic compounds.

Properties of Ionic Solids

The ionic bond is a strong bond, its cohesive energy is large. It is difficult to separate the ions from the condensed aggregate. Their melting points and boiling points are high but have low vapour pressure. There is rupture of bonds on melting although, the ions are grouped together even in melt so much so that they have strong ionic interaction. The structure of solid is very compact as the ions are closely packed.

MOLECULAR CRYSTALS

Substances that exist as aggregates of discrete molecules in the crystalline state are known as molecular crystals. Their melting points are low and are efficiently packed. They are held by Van der Waals forces.

They are either cubic close packed or hexagonally close packed. The interstitial sites or holes are larger in the case of HCP than C.C.P. The tetrahedral hole can accommodate a sphere of radius 0.41 times that of larger sphere while the tetrahedral hole can accommodate a sphere of radius 0.23 times that of larger sphere.

The solid inert gases being spherical in shape form a simple CCP structure. While N_2 and O_2 which are non-sphereoids give a face centred packing CCP. In such crystals there are two types of interatomic distances, one between the atom and its molecular partner and the other between atoms not covalently bonded to one another. These are known as Van der Waal radii. In iodine the internuclear distance is 2.68Å, while the distance between two non-bonded atoms is 4.54Å. The Van der waals radius of iodine if about 2.3Å.

Table 2.4

Element	H	N	O	F	P	S	C	I	As	Se	Br	Sh	Te
Van der Waals Ratio A	1.20	1.50	1.40	1.35	1.90	1.85	1.80	2.00	2.00	1.95	2.2	2.20	2.15

Sulphur, selenium and tellurium have similar structure of long spiral chain packed together with their axis parallel. Phosphorous, arsenic and antimony are dimorphic. They assume both close packed tetrahedral structure and layer type of arrangement.

In BCl_3 or BF_3, the halogen atoms form an HCP structure with boron atoms in the holes (Fig. 2.20).

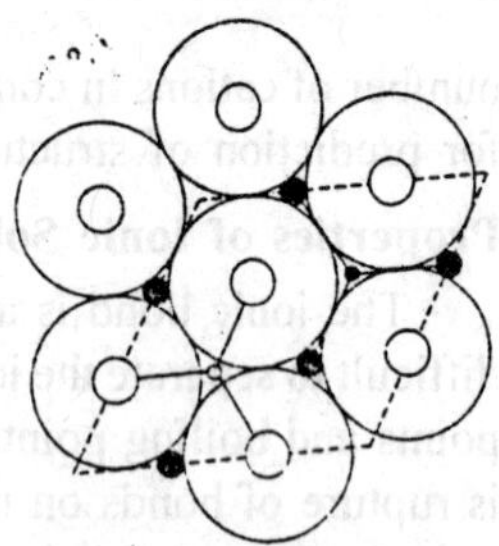

Fig. 2.22

The cohesion energy of a molecular crystal is defined as the enthalpy of the reaction

$$M(g) \xrightarrow{\Delta H} M\ (\text{Crystal}) \quad ...(23)$$

$$\text{or } (H_{fus} + H_{vap}). \quad ...(24)$$

If the close-packed molecules possess a dipole moment then dipole-dipole interaction will take place and this, will contribute to the cohesion energy. Table 2.5 gives various physical properties.

Special properties in the case of H_2O, NH_3 and HF are found due to hydrogen bonding. Hydrogen bonded chains in solid methanol and HF are shown in Fig. 2.23a & b. In the case of those molecules which are non-spherical, like organic compounds, they can still be packed in a compact way. Solid ethane exhibits a cubic structure (Fig. 2.24a & b) but solid ethane and ethylene have HCP with body centred structure (Fig. 2.25).

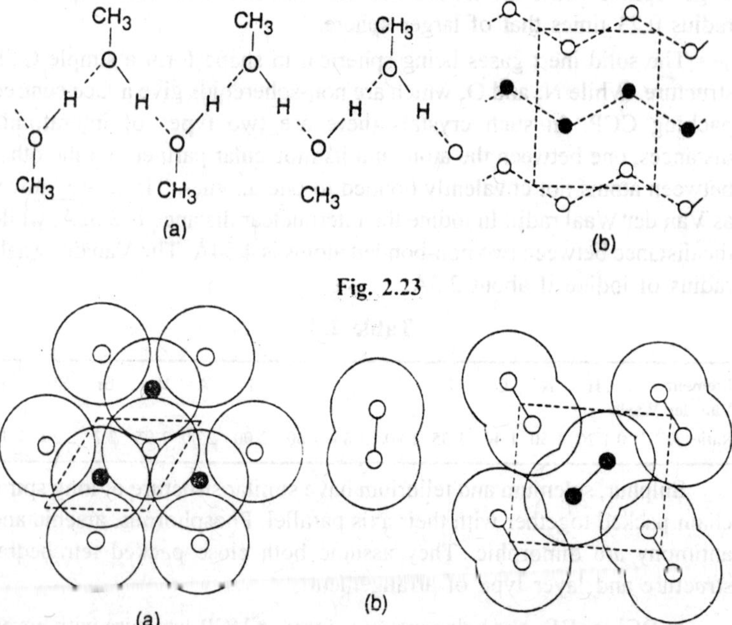

Fig. 2.23

Fig. 2.24

Fig. 2.25

In the case of aromatic compounds some times the planes of molecules are parallel to the crystal axis, *e.g.*, benzene (Fig. 2.26). In the case of naphthalene and phenanthrene, anthracene is better packed than phenanthrene as can be seen from their melting and boiling points.

Table 2.5 : Physical properties of molecules or ions in molecular crystals.

Molecule or ions	*mp °K*	*bp °K*	*ΔH fus K cal/mole*	*ΔH vap K cal/mole*	*μ ion debyes*	*Polarizability α cm³ × 10⁻²⁵*
He	3.5	4.2	0.005	0.02	–	2.0
H_2	55.2	85.2	0.37	1.51	–	–
F	–	–	–	–	–	–
Cl_2	172.0	239.0	1.53	1.53	–	46.1
Br_2	266.0	298.0	2.52	2.52	–	–
Br	–	–	–	–	–	–
H_2	14.0	20.4	0.03	0.22	–	7.9
W_2	63.1	77.3	0.17	1.33	–	17.6
O_2	54.0	90.2	0.11	1.63	–	16.0
S_8	368.0	–	–	3.01	–	–
HF	190.0	190.0	0.09	–	1.98	24.6
HC1	158.0	158.0	0.48	3.86	1.08	26.3
HBr	186.0	186.0	0.56	4.21	0.79	36.1
H_2O	273.0	373.0	1.44	9.72	1.84	–
NH_3	195.0	240.0	1.35	5.58	1.45	–
SO_2	198.0	263.0	1.77	5.96	1.62	–
CO_2	195.0	–	–	6.03	–	26.5
BF_3	145.0	–	–	5.7	–	–
SiF_4	177.0	–	–	6.15	–	–
CH_4	90.7	112.0	0.23	1.96	–	–
SiH_4	89.0	161.0	0.16	2.9	–	–
HCOOH	282.0	373.0	3.03	5.32	1.2	–
CH_3COOH	290.0	391.0	2.80	5.83	1.86	–
CH_3OH	175.0	338.0	0.76	8.94	1.68	–
C_2H_5OH	159.0	352.0	1.20	9.22	1.70	–

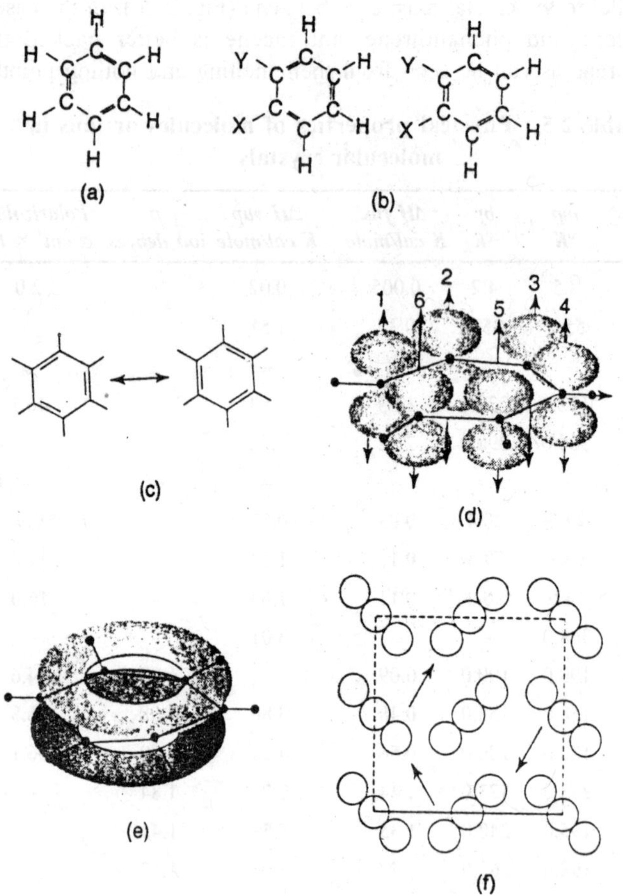

Fig. 2.26

Table 2.6

	Mol wt	*Molar volume*	*mp °C*	*bp°C*
Anthracene	178	142	217	354
Phenanthrene	178	174	996	340

Naphthalene is soluble in benzene but insoluble in water. Anthracene is less soluble than phenanthrene. They are bad conductors of electricity as they do not have free electrons or ions.

SULPHUR

Sulphur acts as a diatomic molecule only at high temperatures in gaseous state. These vapours on cooling give a ring structure of S_8 (Fig. 2.27). It gives two structures rhombic sulphur and β monoclinic sulphur. Rhombic sulphur is stable upto 89° and has a unit cell of 16 S_8 molecules in a ring form with S-S bond length 2.10A and SSS as 103°. β-monoclinic sulphur is stable from 98° to 122°. Sulphur on heating gives liquid which becomes viscous as the temperature is raised to 160°, but gives an amorphous plastic mass on cooling rapidly. This is due to large interaction between sulphur molecules. X-ray studies indicate that it has a helical structure with eight atoms per spiral. The S_8 ring opens and joins to form a chain resulting in a polymer with very small S_8 molecules. Sulphur has been reported to exist in fifteen forms. The solubility of sulphur depends on the arrangement of molecules, CS_2 only dissolves molecules like S_8 whereas long chain molecules are insoluble.

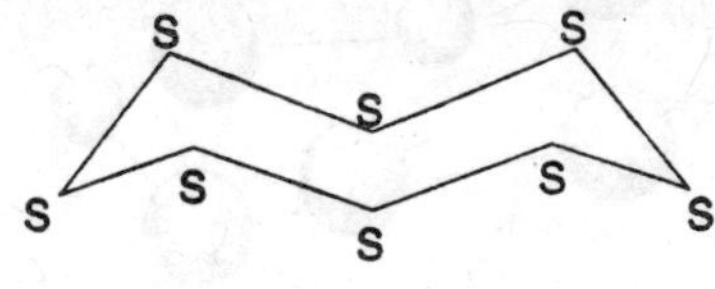

Fig. 2.27

COVALENT SOLIDS

The atoms in a covalent solid occupy such position that the attractive force due to two protons acting on each electron completely counter balances the force of repulsion between two electrons. They equally share electrons with their neighbour. These bonds extend in fixed directions resulting into a giant interlocking structure, *e.g.*, diamond, silicon, germanium, grey tin, etc.

The covalent crystals have high energies (35-150 Kcal/mole), short interatomic distances (1.01-1.6 A), relatively constant angles between successive bonds. These bonds are very strong (high melting point and large heat of sublimation). Their directional character does not lead to maximum filling of space. They are bad conductors of electricity and are chemically inert.

The bonding in these crystals is through sp^3 hybridization. Diamond is insulator, silicon, germanium and grey tin are semiconductors while white tin is metallic (Fig. 2.28a).

Carbon also assumes graphite structure. The bonding in graphite is with sp^2 hybrid orbitals forming a bonds linking the carbon atoms. The MO approach considers the crystal as a large aromatic molecule with π MO'S delocalized parallel to the layers (Fig. 2.28 a & b). Structure of SiO_4 is given in Fig. 2.28c, d & e.

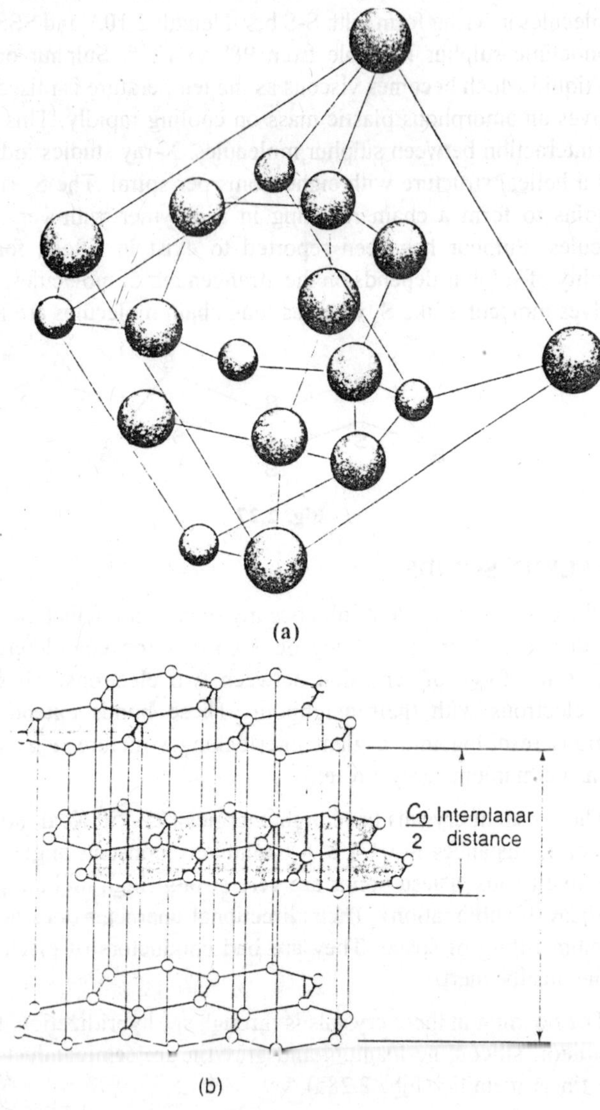

(a)

(b)

Fig. 2.28 (b)

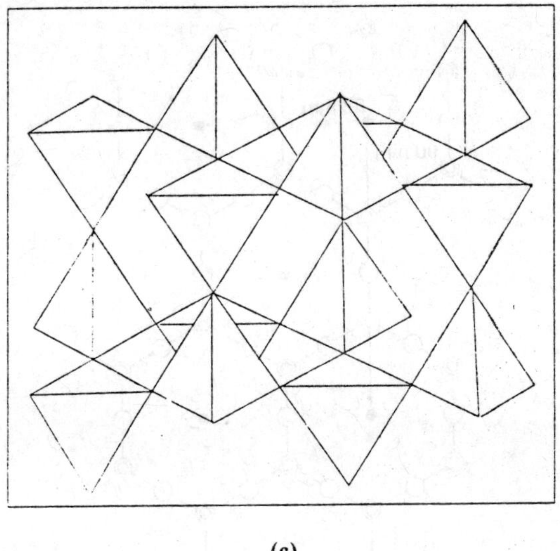

(c)

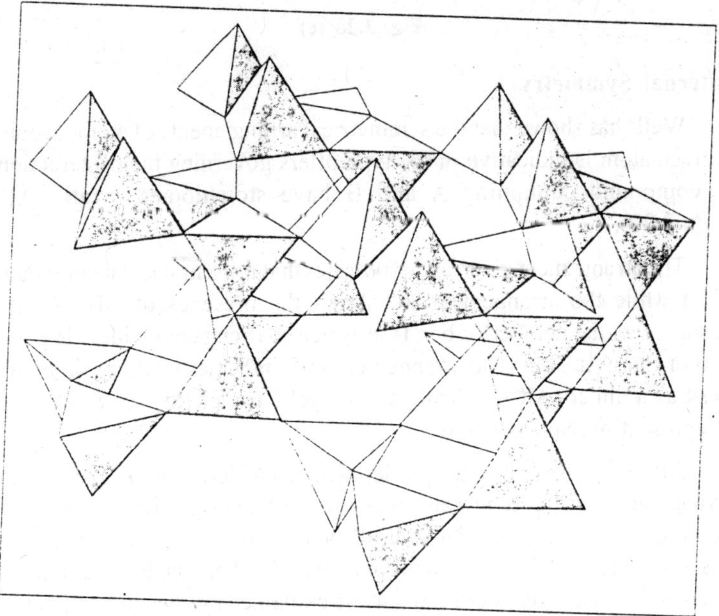

Fig. 2.28 (d)

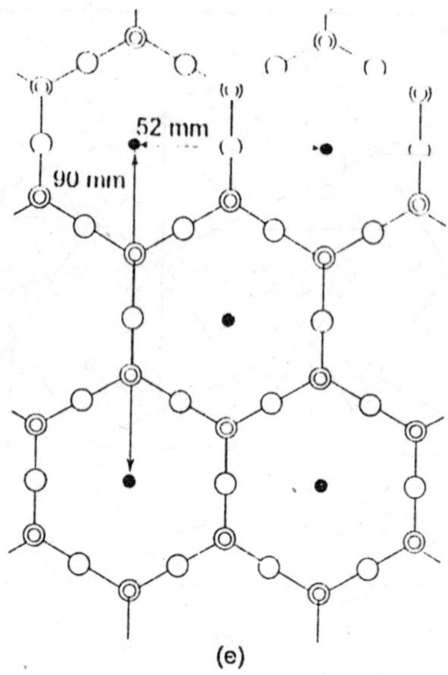

Fig. 2.28 (e)

Internal Symmetry

Wells has shown that the symmetrical arrangements of atoms around a given atom is indicative of the type offers governing their interaction. A compound containing A and B have stoichiometric ratio AB_3 (Fig. 2.29).

The arrangement given in (a) indicates that it is a molecular compound (Fe_3) while the arrangement (b) shows the presence of AB_2 & AB_4 groups, *i.e.*, the number of bonds in which A is engaged with B is either two or four ($AuCl^-_3$). The arrangement in C indicates that the compound exist as a dimer ($AlCl_3$). While d is a cyclic trimer (SO_3 or S_3O_9) c & f represent a chain structure.

All these indicate that a detailed consideration of the spatial arrangement of a system can give a powerful insight of the nature of bonding. Cu, NaCl and diamond all belong to cubic system but NaCl has a heteronuclear linking and in contrast to Cu, diamond has a tetrahedral structure. These differences can only be explained by considering internal symmetry and space groups.

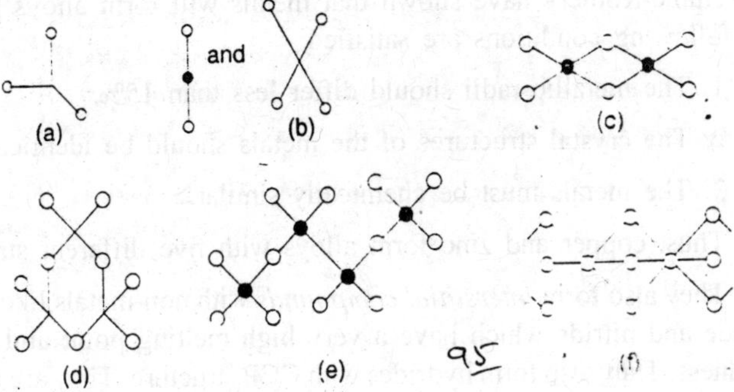

Fig. 2.29

METALS

Those crystalline solids which exhibit high thermal and electrical conductivity, a metallic lusture, a large coefficient of thermal expansion and low ionization energy are known as metals. The structure and properties of metals are given in Table 2.7.

Metallic elements display body-centred, face centred or HCP structures. In metals the radii are found to be functions of coordination number and very often all the neighbours are not equidistant. Goldschmidt deduced that the relative order for different coordination numbers are:

Table 2.7 : Structures of some polymorphic metals

	Structure	*Se angstroms*	*Temperature °K*
α Ca	CCP	3.94	291
γ Ca	BCC	3.87	770
α Fe	BCC	2.48	293
α Fe	CCP	2.57	1189
δ Fe	BCC	2.53	1667
α Sr	CCP	4.29	298
β Sr	HCP	4.31	521
γ Sr	BCC	4.19	887
Li	BCC	3.03	293
Li	HCP	3.11	351

Hume-Rothery have shown that metals will form alloys provided the following conditions are satisfied.

1. The metallic radii should differ less than 15%.
2. The crystal structures of the metals should be identical.
3. The metals must be chemically similar.

Thus, copper and zinc form alloys with five different structures.

They also form *interstitial compounds* with non-metals like carbide, boride and nitride which have a very high melting point and extreme hardness. They also form hydrides with CCP structure. They are regarded as close-packed structures of metals with smaller non-metallic elements located in interstitial sites or holes which are either octahedral .or tetrahedral.

Let us examine the case of lithium on the basis of atomic orbital (a.o.) and molecular orbital (m.o.) bonding.

There are as many m.o.'s as there are a.o's (with each m.o. holding two electrons with opposite spins) and since 2s a.o. of each lithium atom has only one electron, it follows that half the m.o.'s are fully occupied and the other half are vacant. In other words, the 2s-band in lithium is only half-filled. This conclusion remains valid if we consider a three-dimensional array of Li atoms.

This picture of metallic bonding readily explains metallic behaviour. Firstly, since the m.o.'s are spread over the entire crystal, electrons in these m.o.'s are completely delocalized. Secondly, the half-filled nature of the electric field, thus, permitting electrical conduction. The same two factors also account for thermal conductivity. The delocalized electrons absorb and re-emit light, rendering the metal surface shiny and lusturous. Finally, since the bonding electrons are not localized at any particular atom, the lattice can be deformed easily, thus, explaining the malleability and ductility of metals.

Our analysis of lithium band structure applies equally well to other alkali metals. If we apply the same analysis to magnesium which has two electrons in 3s level, the 3s-band would be predicted to be completely filled, thus, making magnesium a non-metal. However, the width of energy bands depends upon crystal structure and the crystal structure of magnesium is such that the 3p-band overlaps with the 2s-band. This means that electrical and thermal conduction can take place and magnesium exhibits metallic behaviour. This example raises the obvious question, "Under what conditions band overlap can be expected?" The answer is also quite obvious. If the orbital interaction (that is orbital overlap) increases, there is greater lowering of energy and also greater overlap of bands involving different orbitals. Orbital irteraction increases with decrease in internuclear distance. One way to shorten the internuclear distance is by applying pressure. Thus, many elements (even non-metallic ones) under high pressure undergo phase transitions and become metallic (conducting).

In fact, calculations show (hat at a pressure of a few hundred thousand atmospheres, the hydrogen atoms should form a cubic close-packed structure which should conduct electricity. Another way that changes in internuclear separation can occur is when the crystal structure of a solid changes.

Thus, grey tin which has relatively open structure (density = 7.29 gm/cc) is a metal. As long as the crystal structure of a metal is not disturbed, its band structure remains unaltered. This is why foreign atoms of the same valency and similar size can readily replace atoms in a metal in arbitrary proportions (forming alloys) without significantly altering the crystal structure or band structure.

BAND MODEL

We shall try to understand it by means of a small ball in a well with steps at different heights. If the ball is rolled on the sides of the well it can either rest at the bottom of well (ground state) or on one of the steps (excited states). If there are more than one balls, the correct situation will be that we start from the bottom up and the number of balls which can be accommodated at each level depends on some parameters that characterize the steps. An atom may be thought of in a well with steps at different heights.

If there are non-interacting atoms, *e.g.*, argon, there would be many such wells separated by interatomic distance whereas in large amount of cases atoms do interact. In that case we have a large number of smaller steps or bands very close to each other between consecutive level. In each band there are as many steps as the atoms (Fig. 2.30a). In a diamond gem of one carat there are about 10^{23} atoms, hence, this will be the order of magnitude of steps with different energies. It is not possible to accommodate the atoms outside the well. If the broadening of band is profound, the forbidden zone between two bands disappears and a wider band is obtained.

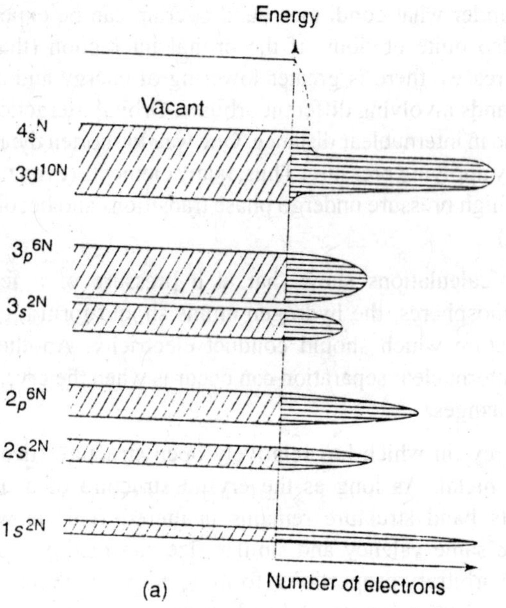
Energy
Vacant
4_s^N
$3d^{10N}$
3_p^{6N}
3_s^{2N}
2_p^{6N}
$2s^{2N}$
$1s^{2N}$
Number of electrons
(a)

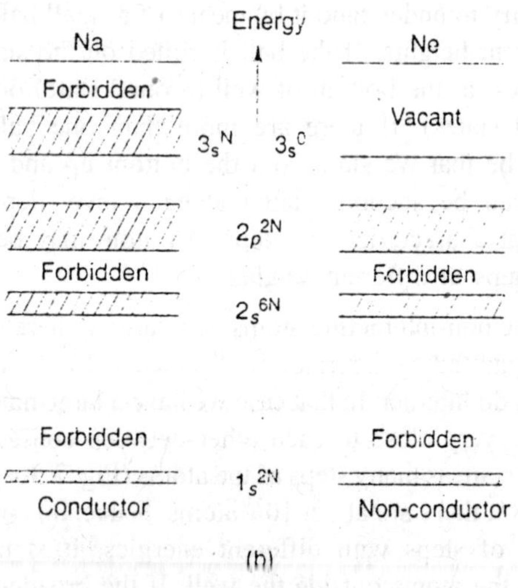
Energy
Na
Ne
Forbidden
Vacant
3_s^N
3_s^0
2_p^{2N}
Forbidden
Forbidden
2_s^{6N}
Forbidden
Forbidden
1_s^{2N}
Conductor
Non-conductor
(b)

Fig. 2.30

Let us imagine some small balls in a container. These will fill the box with a lid in flat condition up to a certain level. If we tilt the container in one direction the balls will roll and occupy positions higher up with respect to the bottom of the box. If the box is completely full, tilting does not produce any moment in the balls. If these balls are electrons, at the bottom of the box, the lower energy limit and the lid, the upper energy limit, it can be seen that if the band is not full, the electrons will have a net movement in the direction of the applied field, *i.e.*, they will show electrical conduction. We can still imagine the balls moving in a full box if they are churned but as many balls will move in one direction as in other. If the band is full, the electrons are not localized in one position, they can move but as many will go in one direction as in other. The net apparent result will be no motion, *i.e.*, no conduction, *e.g.*, diamond.

Since the above model does not depend on the atom, hence, the above model is more universal.

The liquid is empty in non-conductors (diamond, neon, etc.), while in conductors (Cu, Na, etc.) partially filled or overlapping of level takes place (Fig. 2.30b).

IMPERFECTIONS IN SOLIDS

In crystals, the state of complete order and of lowest energy is found at absolute zero of temperature. At any temperature above 0°K there will be some departure from complete order. In principles, any deviation from a completely ordered arrangement in a crystal constitutes native disorder. In addition to native disorder, crystals may have disorder due to the presence of impurities. The term imperfection or defect is generally employed to denote a departure from a perfectly periodic array of atoms in a crystal, immaterial of whether the imperfection is intrinsic or extrinsic in origin. Many a property of a solid such as the electrical conductivity, optical spectra or mechanical strength cannot be explained on the basis of its crystal structure alone. Imperfections may not only change the properties of a crystal, but may also give rise to certain new properties of interest.

The important types of native disorder are those associated with atomic and electronic imperfections. In addition, orientational disorder and notional disorder are also found in certain solids. A typical example of orientational disorder is found in ferromagnetic solids where all the magnetic moments are alligned in magnetic field at low temperatures. At higher temperature, the solid becomes paramagnetic due to the disorientation of the magnetic moments. Motional disorder is exemplified by ammonium halides, where the ammonium ions rotate in phase at low temperatures. At higher temperatures the ions rotate out of phase. In this section, we shall be limiting our discussion mainly to the two more important types of native disorder arising from electronic and atomic imperfections.

Electronic Imperfections

It may appear strange at first sight to consider electrons in solids as imperfections, but we will soon see that such a classification is helpful in understanding several phenomena. In perfect covalent or ionic crystals at 0°K, electrons are present in the fully occupied lowest energy states. The distribution of the valence electrons depends on the nature of the chemical bonds in the case of covalent crystals, while they are mainly concentrated about the electronegative component in ionic crystals. In both these cases electrons do not move under an applied electric field. Above 0°K, some of the electrons may occupy higher energy states depending on the temperature and the energy distribution in the allowed states. Thus, in crystals of pure silicon some electrons are released thermally from the covalent bonds at temperatures above 0°K. These electrons which are free to move in the crystal would be responsible for the electrical conductivity in silicon. The electron deficient bond produced by the removal of an electron is referred to as a positive hole. Holes also give rise to electrical conductivity, but the direction of p motion of the holes in an electric field will be opposite to that of the electrons. Excited electrons and holes in solids are considered to be electronic imperfections. The electrons and holes are generally designated by e and h respectively and their concentrations by n and p.

The concentrations of holes and electrons will be equal in pure crystals like silicon at equilibrium. Electrons and holes can be preferentially produced in covalent crystals like germanium or silicon by adding suitable impurities. Impurities with excess valence electrons (P or As) give rise to extra electrons in germanium and silicon; impurities like Ga or Al with one electron less than Ge or Si produce holes.

In most ionic solids electrons are localized on a particular ion (unlike in metals where electrons are delocalized). The concept of electrons and holes can be employed in ionic crystals just as covalent solids. This is best understood by considering an ionic crystal where cations of two different valence states can be present. Consider a solid containing mainly ferrous ions. If we introduce a small number of Fe^{+3} ions in this solid, it is equivalent to introducing holes. This is because, Fe^{+3} ions have one electron less than the preponderant Fe^{+2} ions in the solids. In such a solid, the electrons and holes spend appreciable time at cation sites before 'hopping' to the adjacent sites. The hopping process may be visualized in terms of the change, $Fe^{+2}\ Fe^{+2}\ Fe^{+3} \rightarrow Fe^{+2}\ Fe^{+3}\ Fe^{+2}\ Fe^{+2}$. The hopping process is responsible for the electrical conduction in such solids.

Atomic Imperfections

If imperfections in a crystal are caused by departure from the periodic arrangement in the vicinity of an atom or a group of atoms, the imperfections are called point defects. If the deviations from periodicity extend over microscopic regions of the crystal, they are referred to as lattice imperfections. Lattice imperfections may extend along lines (line defects) or surfaces (plane defects). Line defects are also called dislocations. The various types of atomic imperfections in crystals are listed in Table 2.8.

Table 2.8 : Atomic imperfection in crystals

Imperfection	*Nature of imperfection*
Point defects	
Schottky defect	Atom missing from the normal lattice site creating a vacancy.
Interstitial	Atom in a normally vacant interstitial site.
Anti structure	Atoms misplaced wrongly in site assigned for some other atoms.
Frenkel defect	Atom in the lattice site displaced to an interstitial site creating a vacancy.
Line defects	
Edge dislocation	Dislocation line marks the edge of an extra plane of atoms which is inserted part way in a crystal.
Screw dislocation	Displacement of atoms in one part of a crystal relative to the rest of the crystal causes a spiral around the dislocation line.
Plane defect	
Grain boundary	Boundary between two crystals in a polycrystalline solid.
Stacking faults	Boundary between two layers with different stacking sequences in close-packed structures.
Shear structures	Structures where excess atoms are present in folded planes connecting slabs of the crystal of normal structure.

Point Defects

When an atom is missing from its normal lattice site a lattice vacancy (Schottky defect) is created. In the case of silver halide, the silver ion is trapped in the interstitial space and thus, creating a hole at some other place (Fig. 2.31a).

In the case of alkali halide crystals there are electron traps known as V and F centres. Heating the sodium chloride crystal in vapour of sodium generates F centres, while in the presence of chlorine vapour a υ centre is generated.

An F centre is an electron in a halide ion vacancy and the defect is known as Schottky defect (Fig. 2.31b).

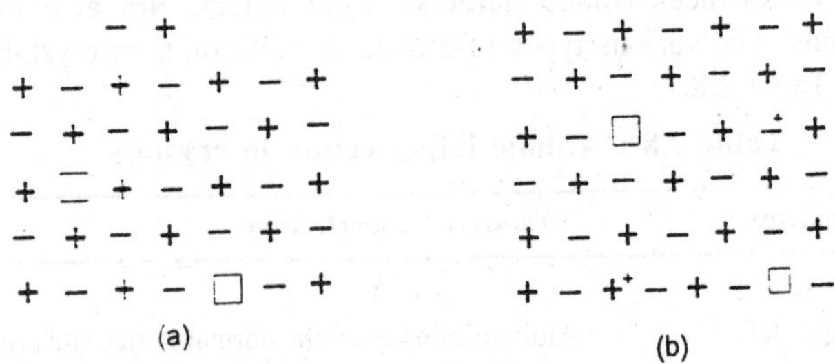

Fig. 2.31

In stoichiometric ionic crystals, a vacancy of one ion is accompanied by the creation of a vacancy of the oppositely charged ion in order to maintain etectroneutrality. Thus, equal numbers of cation and anion vacancies are found in alkali halides. In NaCl, there are approximately 10^6 Schottky pairs per cc at room temperature; in one cc there are about 10^{22} ions and therefore, there will be one Schottky defect for 10^{16} ions. Presence of a large number of Schottky defects in a crystal lowers its density markedly. The pyknometric density of vanadium monoxide, VO, is 5.6 g/cc while the x-ray density is 6.5 g/cc. Obviously, there is about 15% disorder of the Schottky kind in VO.

Atoms (or ions) which occupy the normally vacant interstitial sites in a crystal are called interstitials. The important factor determining the formation of interstitials is the size of the atom. There is another kind of point defect called the antistructure disorder, where atoms occupy sites which in the ideal crystal are assigned to atoms of a different kind. In a crystal containing atoms X and Y, some of the X atoms occupy Y sites and an equal number of Y atoms occupy X sites, if there is antistructure disorder in the crystal. Antistructure disorder is commonly found in alloys, particularly at higher temperatures, but is rare in ionic crystals since it would bright like ions next to each other.

In addition to the three basic types of point defects (Schottky, Interstitial and Antistructure) discussed above, three hybrid type of defects can arise from combinations of any two of the basic types as shown below. Of these three combinations the Frenkel defects are more common in ionic solids. In pure alkali halides one does not encounter Frenkel defects since the ions cannot get into the interstitial sites. Frenkel defects are however, found in silver halides where the Ag^+ ions are considerably smaller than the halide ions. Unlike Schottky defects, Frenkel defects do not change the density of the solid. In many real systems various types of point imperfections may occur simultaneously. Thus, in NiAl, Schottky and antistructure disorder are known to be present; in AgBr Schottky and Frenkel defects coexist.

There is a large variety of non-stoichiometric inorganic solids which contain an excess or deficiency of one of the components. Such solids showing deviations from the ideal stoichiometric composition form an important group of defect solids. There are some solids which are difficult to prepare in the stoichiometric composition; thus, the ideal composition in compounds like FeO is metastable (normally we get $FeO_{0.95}O$). The most stable compositions of praseodymium and terbium oxides under ordinary conditions are Pr_6O_{11} and Tb_4O_7 respectively. Similarly, many of the II-VI group compounds such as ZnO and CdS are difficult to prepare in their stoichiometric compositions. In fact, compounds become more interesting to a solid state chemist when there is non-stochiometry. In certain solids, gross variations in composition are also accompanied by changes in crystal structure, while in some others marked deviations from the stoichiometric composition may occur without any change in crystal structure. Thus, VO and TiO can be prepared over a range of compositions (VO_x or TiO_x; x = 0.6 –1.3) with NaCI structure.

Zinc oxide loses oxygen reversibly at high temperatures and turns yellow in colour. The excess metal is accommodated interstitially, giving rise to electrons trapped in the neighbourhood. The enhanced electrical, conductivity of the non-stoichiometric ZnO arises from these electrons (n-type conduction). In FeO, the non-stoichiometry arises from an excess of anions and the electrical conductivity is due to positive holes (p-type conduction). In such oxides the deviation from stoichiometry is accommodated by the change in the valence state of the cation; the cation valency increases if there is anion excess. These two behaviours of oxides can be understood in terms of the basic chemical reactions of the type given below:

Oxygen (anion) excess: $1/2\ O_2(g) = O^{-2}(s)$ + cation vacancy + 2 h (p-type)

Oxygen (anion) deficiency: $O^{-2}(S) = 1/2\ O_2g)$ + anion vacancy + 2e (n-type)

Anion vacancies in alkali halides are readily produced by heating the alkali halide crystals t in an atmosphere of the alkali metal vapour. When the metal atoms deposit on the surface of the alkali halide crystal, halide ions diffuse to the surface and combine with the metal atoms. The electrons produced by the ionization of the alkali metal atoms then diffuse into the crystal and combine with a negative ion vacancy, electrons trapped in anion vacancies (referred to as F centers) give rise to interesting optical and electrical properties. Thus, excess potassium in KCl makes the crystal appear violet, and excess lithium in LiCI makes it pink. Alkali halides with holes trapped in cation vacancies (V centers) can be produced by heating the alkali halides in an atmosphere of the halogen. Such colour centers can be produced by the irradiation of alkali halides by high energy radiation like x-rays or γ-rays.

Till now the discussion of point defects has been limited to systems where there were no foreign atoms or impurities. Foreign atoms can occupy either interstitial or substitutional sites in a host crystal. The electronic structure of the impurity is of importance in the case of substitutional solid solutions while the size of the impurity atom determines the formation of interstitial solid solutions. Many of the impurity systems provide new and interesting properties of great importance; the most well-known of such solid solutions are those of Group III or Group V impurities with Group IV elements like Ge or Si. Group III elements like Ca or Al and Group V elements like P or As enter Ge or Si lattice substitutionally. The Group V elements will have one excess valence electron after forming the four covalent bonds normally formed by Group IV elements. The excess electrons may be released to move through the host structure by thermal energies to give rise to n-type conduction. A Group III element which has only three valence electrons creates one electron deficient bond giving rise to a hole which can move in the crystal carrying an effective positive charge. The electrical conductivity will therefore, be of p-type. In solid solutions of Group IV elements with either Group V or III elements, the overall charge neutrality is maintained. The impurities in these cases are ionized: imperfections that ionize releasing electrons (*e.g.*, P or As in Ge or Si), are called donors while those which ionize by accepting electrons (*e.g.*, Ga or Al in Ge or Si) are called acceptors.

The magnitude of energy required to ionize impurity atoms is generally small and in germanium at ordinary temperatures (0.025 eV) nearly all the donor and acceptor levels are ionized. The ionization energy decreases with the increase in impurity concentration and if the concentration is sufficiently high, a large fraction of ionized donors and acceptors is found even at low temperatures. The extra electron of the donor atom may be visualized as moving in orbits in the field of a positive charge. The picture is somewhat similar to that of the hydrogen atom, the main difference being that the electron and the positive charge are in a medium of high dielectric constant. Further, the radius of the orbit is much larger covering several atomic distances.

A common method of introducing defects in ionic solids is by adding impurity ions. If the impurity ions are in different valence state from that of the host ions, vacancies are created. Literature abounds in examples where interstitials or vacancies are produced by addition of impurities to ionic solids. For example, addition of $CdCl_2$ to AgCl or $SrCl_2$ to Nad yields solid solutions where the divalent cations occupy the Ag^+ or Na^+ sites and produce cation vacancies equal in number to that of the divalent ions (Fig. 2.32). If one adds a molecule of a rare earth fluoride, MF_3 to CaF_2, the type of the concentration of defects produced will depend on the valence state of the rare earth ion in the host lattice. If the rare earth ion is in the +3 state, one F^- interstitial will be formed beside a new cation site; if the rare earth ion is in the +2 state there will be no interstitial F^- ion. The valence state and the defect structure in such systems is best studied by spectroscopic methods.

If the ions in a host crystal can assume two or more valence states, addition of an impurity ion in a valence state different from the normal valence state of the ions in the host crystal produces changes in the valence state of a controlled number of host ions. Such controlled valency systems show interesting semiconducting properties. A good example of a controlled valency semiconductor is NiO doped with Li_2O where the number of N^{+3} ions in the system is determined by the number of Li^+ ions. The composition of the resulting material may be written as Ni^{+2}_{1-2y} Ni^{+3}_{y} $Li^{+}_{y}O$. With 10% atoms of Li. This material is black in colour and exhibits high electrical conductivity (p-type); NiO is pale green in colour and is an insulator.

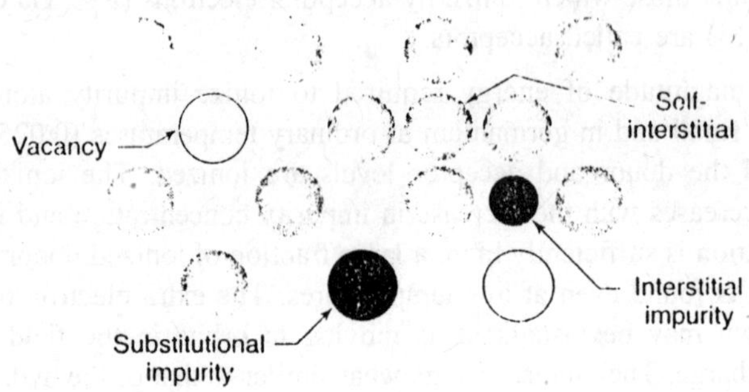

Fig. 2.32

The most important criterion in the defect chemistry of ionic crystals is the principle of electroneutrality or charge compensation. The creation of vacancies or new valence states of ions in crystals by the addition of impurities is determined by this simple principle.

SEMICONDUCTORS

Solids, containing ions or ion vacancies which move under the influence of applied field are known as semiconductors. Their conductivity is intermediate between metals and insulators and is of the order 10^2 to 10^{-10} mho cm^{-1}. The electrical conductivity in metals decreases with the increase in temperature but in semi-conductors it increases with the increase in temperature. Their conductivity is either intrinsic or due to impurities (Fig. 2.33a, b, c).

Unlike metals, the conductivity of semiconductors and insulators are mainly determined by the impurities and defects. Electrons and holes produced by the ionization of defects contribute to the electronic conduction in these solids. The behaviour of insulators and semi-conductors can be explained in terms of the band model. If the band structure of an element is such that the fully occupied band is separated by a large energy gap from the vacant band, then conduction is not possible. This situation is found in good insulators like diamond or MnO. On the other hand, if the energy gap is small, promotion of electrons can take place by raising the temperature; higher the temperature greater

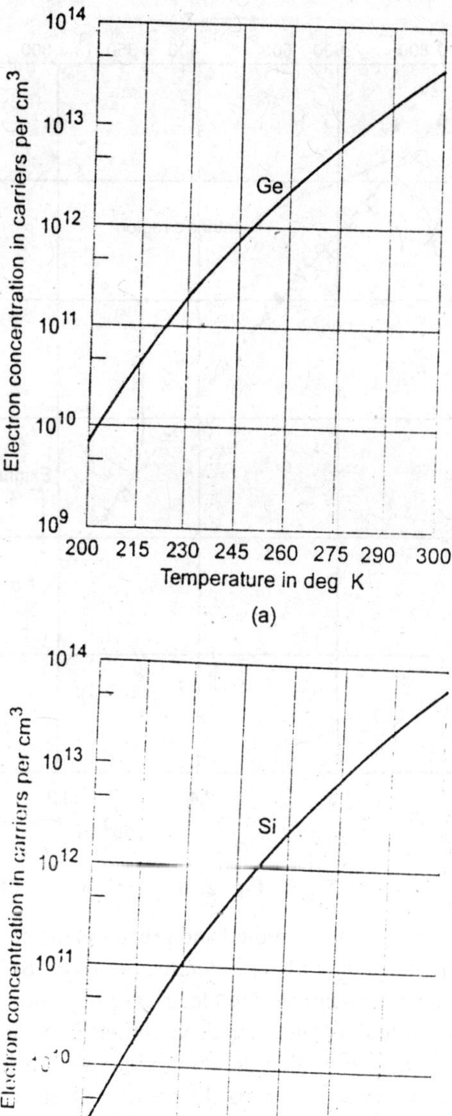

10^{14}
10^{13}
10^{12}
10^{11}
10^{10}
10^{9}
Electron concentration in carriers per cm^3
Ge
200 215 230 245 260 275 290 300
Temperature in deg K
(a)
10^{14}
10^{13}
10^{12}
10^{11}
10^{9}
Electron concentration in carriers per cm^3
Si
275 300 325 350 375 400 425 450
Temperature in deg K
(b)

Fig. 2.33

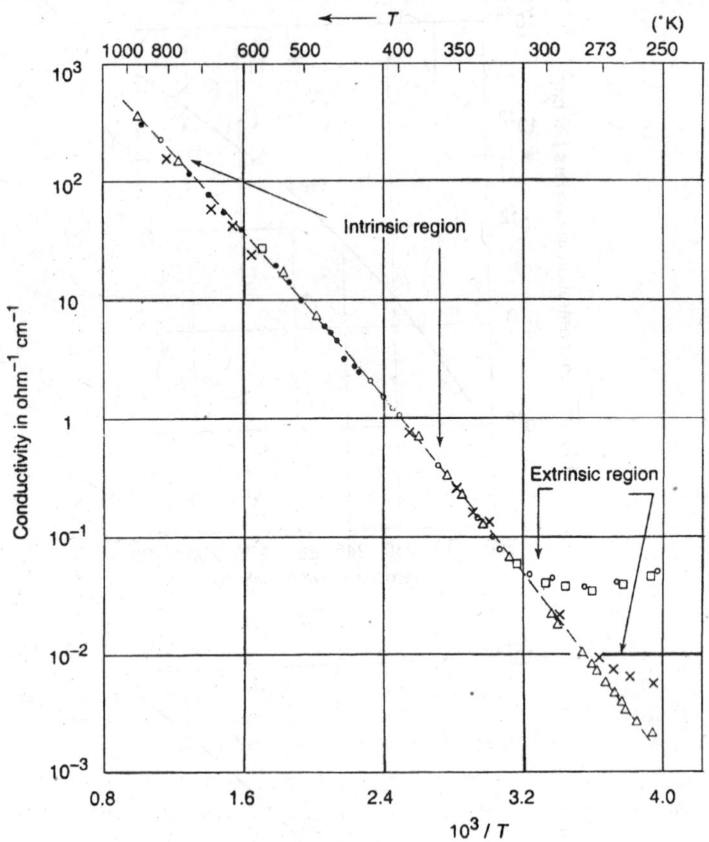

Fig. 2.33

is the number of electrons promoted and greater is the conductivity. This is characteristic behaviour of semi-conductors (*e.g.*, SnO_2; $SrTiO_3$). In Table 2.9 we have summarized the electrical properties of some oxides of transition metals to illustrate the wide variations found in such materials. It is particularly interesting that the monoxides all of which possess the NaCI structure show such marked differences in electrical properties; this is obviously related to electronic configurations of the transition metal ions.

In addition to electronic semiconductors where the band picture is applicable, there are many materials where the semiconduction is due to the hopping of electrons from one site to another. Examples of these are oxides where cations of more than one valence state are present as in Pr_6On, NiO doped with LiO, etc.

Table 2.9 : Electrical properties of typical transition metal oxides

TiO(M)	VO(M)	CrO(?)	MnO(I)	FeO(s)	CoO(s)	NiO(s)	CuO(s)
Ti_2O_3(M-S)	V_2O_3(M-S)	Cr_2O_3(s)	Mn_2O_3(s)	Fe_2O_3(s)	CO_3O_4(S)	–	Cu_2O(s)
TiO_2(s)	VO_2(M-S)	CrO_2(M)	MnO_2(D)	Fe_3O_4(D)			
	V_2O_5(S)		Mn_3O_4(s)				

M = Metal; I = Insulator; S = Semiconductor; M-S = Shows a transition from metal to semiconductor behaviour at a certain temperature; ? = not known; D = Degenerate semiconductor (high conductivity with little or no variation with temperature).

Ionic conduction in a solid involves migration of atoms, ions or other charged species under applied field and it is generally useful to study the diffusion (migration due to a concentration gradient) of these species in the solid. Conductivity measurements in ionic solids like NaCI give valuable information on the energy required for the creation of defects, as well as their migration energies.

INSULATOR

The number of electrons promoted will increase exponentially with temperature. Thus, an insulator can in principle be made more conductive by thermal excitation of electrons but if the energy gap is so large that the number of electrons excited even at high temperature is negligible, the material is classified as insulator. If the energy gap is very small there will be an effective thermal excitation at low temperatures (Fig. 2.32a)(i).

THERMAL EFFECT

The effective average energy of the electron in thermal equilibrium with the surroundings is called Fermi energy. In the case of metals it is the energy of the electrons that can be thermally promoted to a conduction state. At absolute zero it is the energy of the most energetic electron. In the case of insulators it lies about the topmost valence band.

The thermal excitation of the electrons in an insulator provides a few conduction electrons in the upper band and in the lower band an equal number of mobile holes are generated. The energy of these current carrying holes and electrons lies in the energy gap that separates them.

When the thermally produced electrons and holes are mainly responsible for the conductivity of a semiconductor (generally at higher temperatures) rather than the electrons and holes produced by donors and acceptors (as in doped Ge or Si), the region is called the intrinsic region (Fig. 2.33c). At lower temperatures where the conductivity is mainly determined by the concentration of donors and acceptors, the region is called extrinsic. We should recapitulate here that pure germanium or silicon by itself would be an insulator; only by doping with group III or V impurity can we increase the conductivity at ordinary temperatures (Fig. 2.34b & c).

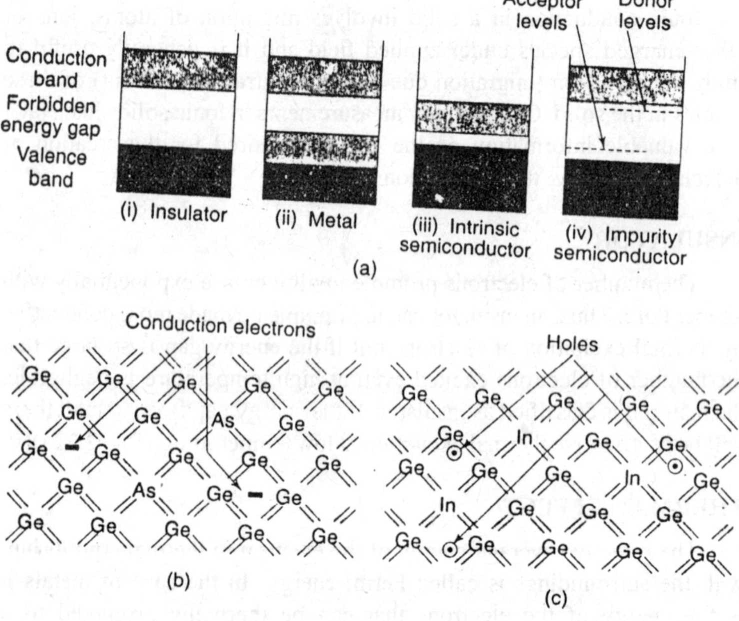

Fig. 2.34

The intrinsic conductivity in a semiconductor is due to the thermal excitation of electrons from valence band into empty states of the conduction band with energy gap of the order of RT or 600 cal mole^{-1}. As this energy gap approaches zero, valence and conduction band merge and the conduction becomes metallic Fig. 2.34a(ii).

Impurity conductivity is either due to foreign atoms or lattice imperfections.

If pure Si or Ge is exposed to the vapours of P or As, they are held into the lattice of Si or Ge. The extra electron associated with each foreign atom remains near by and seeks a state of low energy in the range of energies forbidden to electrons in pure Si or Ge. Since they carry negative charge they are called n-type of semiconductor (Fig. 2.35). In the case of Si, the minimum energy required to transfer an electron from an impurity centre to conduction band is 1.84 Kcal J mole^{-1}.

If B, A*l*, or Ga are held into the lattice of Si or Ge each foreign atom will be associated by electron valencies or holes. These electrons are mobile. When a hole is filled by an electron, a hole is created in some other position. So in effect these holes move contributing to the conductivity of the solid. Since they are positive charge carriers, they are called p-type of semiconductors.

These semiconductors are used as solid state electronic component called transistor.

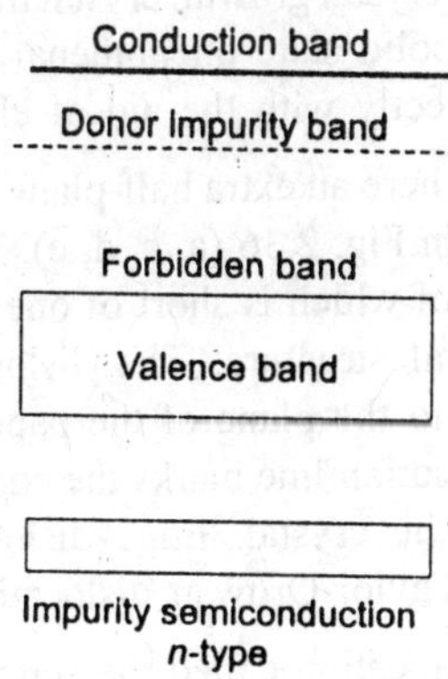

Fig. 2.35

ELECTRICAL PROPERTIES

Based on their electrical conductivity, solids can be broadly classified into three types: metals, insulators and semiconductors (Fig. 2.34a).

Conductivity of solids varies anywhere from 10^8 ohm^{-1} cm^{-1} in metals to 10^{-22} ohm^{-1} cm^{-1} in insulators. Electrical conductivity of solids may arise through the motion of electrons and holes (electronic conductivity) or other charged imperfections and ions (ionic conductivity). Substances like pure alkali halides, where the conduction is through ions only are generally insulators since the mobilities of ions are very much lower than those of holes or electrons. However, presence of vacancies or other imperfections could markedly increase the conductivity of ionic solids.

At laboratory temperatures, conductivity of metals is nearly independent of impurities and lattice defects. The electron concentration is mainly determined by the nature of the metal (number of valence electrons in the metal) and the mobility by the lattice vibrations. At low temperatures, however, the lattice vibrations are not limiting and the conductivity should be infinite. However, this is not true because of the presence of lattice imperfections and impurities. It is thus, convenient to use the resistance ratio p 300°K/ρ. 4.2°K as a measure of purity of the metal.

LINE DEFECTS

Although the concept of dislocations was first introduced primarily to explain deformation in metals, it is now established that dislocations play an important role in crystal growth, crystal transformations, electrical conductivity and other solid state phenomena. Dislocations in crystals have been observed directly with the aid of electron microscopes.

Edge dislocations where an extra half-plane of atoms is inserted into the lattice is illustrated in Fig. 2.36 (a, c, d, e). The dislocation consists of a line of atoms each of which is short of one coordinating atom than prescribed by the crystal structure. The dislocation extends along a direction perpendicular to the plane of the paper and is designated by the symbol ⊥. The dislocation line marks the edge of the plane of atoms inserted part-way into the crystal. Edge dislocations are also called Taylor dislocations or Taylor-Orowan dislocations.

Another kind of dislocation called the screw dislocation described by Burgers (also known as Burgers dislocation) consist of a line of atoms with distorted coordination polyhedra. This is caused by a displacement of atoms in one part of the crystal with respect to the rest of the crystal, the displacement direction being parallel to the dislocation line; as shown in Fig. 2.37b & c. The row of atoms marks the termination of the displacement. The polyhedron is not regular in the disturbed region and

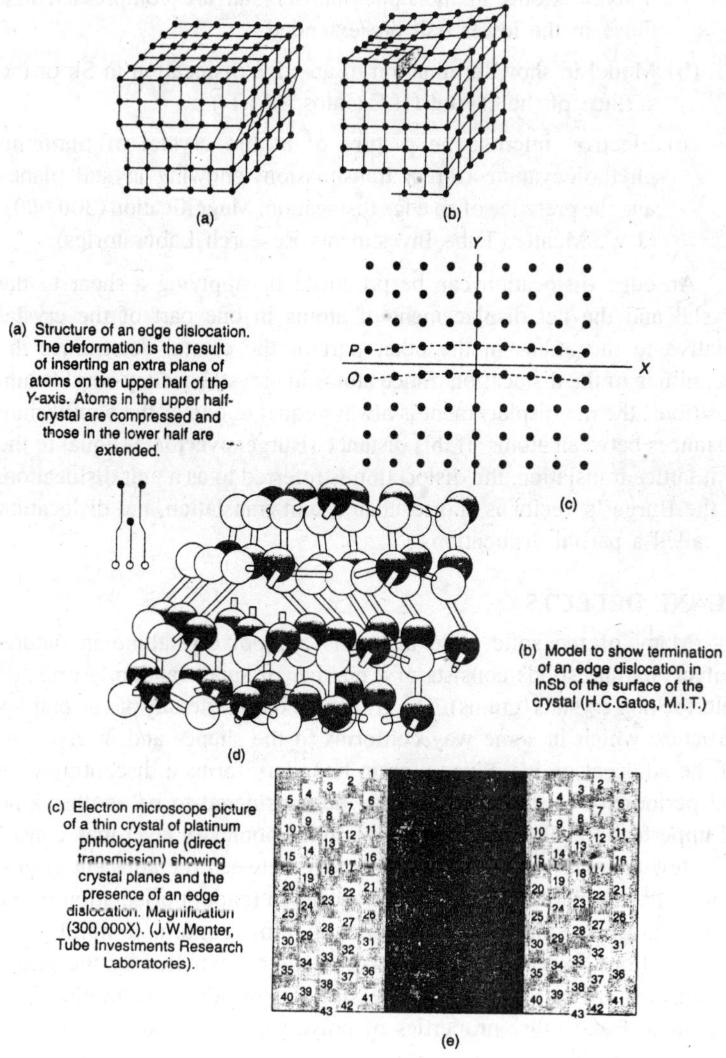

(a) Structure of an edge dislocation. The deformation is the result of inserting an extra plane of atoms on the upper half of the Y-axis. Atoms in the upper half-crystal are compressed and those in the lower half are extended.

(b) Model to show termination of an edge dislocation in InSb of the surface of the crystal (H.C.Gatos, M.I.T.)

(c) Electron microscope picture of a thin crystal of platinum phtholocyanine (direct transmission) showing crystal planes and the presence of an edge dislocation. Magnification (300,000X). (J.W.Menter, Tube Investments Research Laboratories).

Fig. 2.36

the distortic decreases progressively with increasing distance from the dislocation line.

(a) *Structure of an edge dislocation* : The deformation Is the result of inserting an extra plane of atoms on the upper half of the

V-axis. Atoms in the upper half-crystal are compressed and those in the lower half are extended.

(b) Model to show termination of an edge dislocation in Sb of the surface of the crystal (H.C.Gatos, M.I.T.)

(c) Electron microscope picture of a thin crystal of platinum phytholocyanine (direct transmission) showing crystal planes and the presence of an edge dislocation. Magnification (300.000). (J.W. Menter, Tube Investments Research Laboratories).

An edge dislocation can be produced by applying a shear to the crystal and the net displacement of atoms in one part of the crystal relative to the atoms in the other part of the crystal determines the magnitude of the dislocation. Since atoms in acrystal occupy equilibrium positions, the net displacement is always equal to one of the equilibrium distances between atoms. If this distance (Burgers vector) is equal to the unit lattice translation, the dislocation is referred to as a unit dislocation. If the Burger's vector is shorter than a unit translation, the dislocation is called a partial dislocation.

PLANE DEFECTS

Many of the solid state materials are polycrystalline in nature. Polycrystalline solids consists of a number of small randomly-oriented interlocking crystals (grains). The grain boundary generally has a complex structure which in some way conforms to the shapes and orientations of the adjacent grains. Since a grain boundary forms a discontinuity in the periodicity of the lattice, it may be considered to be another kind of imperfection. The disturbed region in the boundary layer may consist of a few atom layers and acts as a bridge between the adjacent grains. These disturbances are generally large due to random orientation of the initial grains. Since grain boundaries may contain a number of imperfections like impurities, interstitials, etc. (expelled by the grains during their growth), it often becomes difficult to understand the electrical, mechanical and other properties of polycrystalline solids.

Another important kind of plane defect found in crystals arises from stacking faults. In the cubic close-packed structures, we have the sequence ... ABCABC ..., while in the hexagonal case we have the sequence ... ABAB ... A fault in a stacking sequence as in ABABBCBCBC or ABCABCBCABCABC, occurs when the packing is cubic instead of hexagonal or vice versa. The sequence in the planes before and after the stacking fault would, however, be normal.

PROPERTIES OF SOLIDS

There is a close relationship between the physical properties of a solid and its structure and chemical composition. We shall now examine a few of the important properties of solids to illustrate the wide variety of possibilities. Some of these properties are exploited for various new innovations in electronic and magnetic devices such as transistors, computers, telephones, etc. The information explosion in inorganic and organic solid state materials has been so great in recent years that it would be difficult to list representative examples of all classes of materials in an elementary chemistry text of this sort.

Magnetic Properties

Materials can be divided into different classes depending on their response tq, magnetic fields. Diamagnetic materials (like NaCl or ZrO_2) are weakly repelled by magnetic fields. In paramagnetic materials there are permanent magnetic dipoles due to the presence of atoms, ions or molecules with unpaired electrons (*e.g.*, O_2, NO, Na atoms, Ti_2O_3, VO_2) and these materials are attracted by magnetic fields; they, however, lose their magnetism in the absence of a magnetic field. Solids like TiO_2 which should be diamagnetic often exhibit paramagnetism due to the presence of slight non-stoichiometry. Unlike paramagnetic materials, ferromagnetic materials show permanent magnetism even after the magnetic field is removed (*e.g.*, Fe, Cu_2Mn, Sn, EuO, CrO_2). The explanation for the large magnetization in these materials is that there are domains of magnetization (consisting of about a million atoms or ions), all of which cooperatively direct their magnetic moments (spans) in the same direction. Once such a material is magnetized (with all domains oriented), it remains permanently that way. Iron, cobalt, and nickel are the only three elements which show ferromagnetism at room temperature. Manganese has most of the properties required for ferromagnetism, but the metal atoms are too close; however, addition of copper to manganese enhances the spacing and the alloy is ferromagnetic.

We have just learnt that a spontaneous alignment of magnetic moments in the same direction gives rise to ferromagnetism. If the alignment of moments is in a compensatory way so as to give zero net moment, then we get antiferromagnetism in the material (*e.g.*, MnO). Ferrimagnetism is found when the moments are aligned in parallel and antiparallel directions in unequal numbers resulting in a net moment (*e.g.*, Fe_3O_4, ferrites of the formula $M^{+2}Fe_2O_4$ (M = Mg; Cu, 3n, etc.). In Fig. 2.35a, b & c we have shown the spin alignments in antiferromagnetic MnO and ferrimagnetic Fe_3O_4. All these magnetically ordered solids transform to the paramagnetic state at a higher temperature due to the randomization of spins. For example V_2O_3 and NiO transform from the anti-ferromagnetic phase to the paramagnetic phase at 150°K and 523°K respectively: Ferromagnetic CrO_2 becomes paramagnetic at 320°K; a similar transition in ferrimagnetic Fe_3O_4 is at 850°K. Magnetic properties of a few transitional metal oxides.

Table 2.10 : Magnetic properties of typical transition metal oxides.

TiO(p)	VO(p)	CrO(?)	MnO(af)	FeO(af)	CoO(af)	NiO(af)CuO(p)
Ti_2O_3(p)	V_2O_3(af)	Cr_2O_3(af)	Mn_2O_3(af)	Fe_2O_3(afw)	CO_3O_4(fe)	–Cu_2O(d)
TiO_2(d)	VO_2(p)		MnO_2(af)	Fe_3O_4(af)		
	V_2O_5(d)		Mn_3O_4(p)			

p = paramagnetic; ? = not known; af = antiferromagnetic; afw = antiferromagnetic with parasitic ferromagnetism; fe = temmagnetic; f = ferromagnetic.

Dielectric Properties

In insulators, electrons are closely bound to individual atoms or ions and they do not generally migrate under an applied electric field. However, dipoles are created by a shift in the charge resulting in polarization.

(i) These dipoles may align themselves in an ordered manner such that there is net dipole moment in the crystals or (ii) they may align themselves in such a manner that the dipole moments cancel each other. (iii) It is also possible that there are no polar species (dipoles) in the crystal, but only ions are present. Crystals where the situation (i) is found, exhibit piezoelectricity. When such a crystal is deformed by mechanical stress, electricity is produced due to the asymmetric displacement of ions or conversely if an electric field is applied to the crystal, there will be atomic displacements causing mechanical strain. Thus, a piezoelectric crystal acts as a mechanical-electrical transducer. Some of the polar crystals, when heated, produce a small electric current or pyroelectricity due to the asymmetric variation in interatomic distances.

In some of the piezoelectric crystals, the dipoles are permanently lined up (spontaneously polarized) even in the absence of an electric field and the direction of polarization can be changed by applying an electric field. This phenomenon is called ferroelectricity by analogy with ferromagnetism. Barium titanate ($BaTO_3$), sodium-potassium tartrate (Rochelle salt), sodium nitrite ($NaNO_2$) and potassium dihydrogen phosphate (KH_2PO_4) are typical ferroelectric solids. The occurrence of ferroelectricity undoubtedly depends on the structure and only certain geometric arrangements in the crystal can exhibit this phenomenon. If the dipoles in alternate polyhedra point up and down, there will be no net dipole moment and the crystal is said to be antiferroelectric. Lead zirconate ($PbZrO_3$) is a typical antiferroelectric. Both ferroelectric and antiferroelectric crystals transform at some temperature to the paraelectric phase where the dipoles are randomly oriented.

NON-CRYSTALLINE SOLID STATE

The liquid phase is the state of substances at temperature above their melting points. This includes all amorphous substances (silicate glass, rosin, etc.) Since silicate glass has no crystalline lattice, all amorphous solids are termed as glassy or glasses. Therefore, the two phase states of solids are crystalline and amorphous (glass).

The glass transition is the passage of a molten liquid into the solid state without changing the phase, *i.e.*, with the retention of short range order. When the liquid is cooled at a fast rate, its coefficient of viscosity increases and its energy of thermal motion decreases. The molecules remain in the random disorder characteristic of a liquid. Since it involves the breaking and reforming of strong covalent bonds, there is only a short range order and a long-range disorder ultimately at a certain temperature at which B the coefficient of viscosity approaches 10^{13} poises which correspond to viscosity coefficient of solids. So the super-cooled liquid solidifies and the substance passes into amorphous (glassy) state.

There are two types of substances that can form glasses.

Glass

Substances, whose liquids consist of long molecules, that become very much entangled, form glasses. Organic polymers and polymeric form of sulphur, selenium, belong to this class.

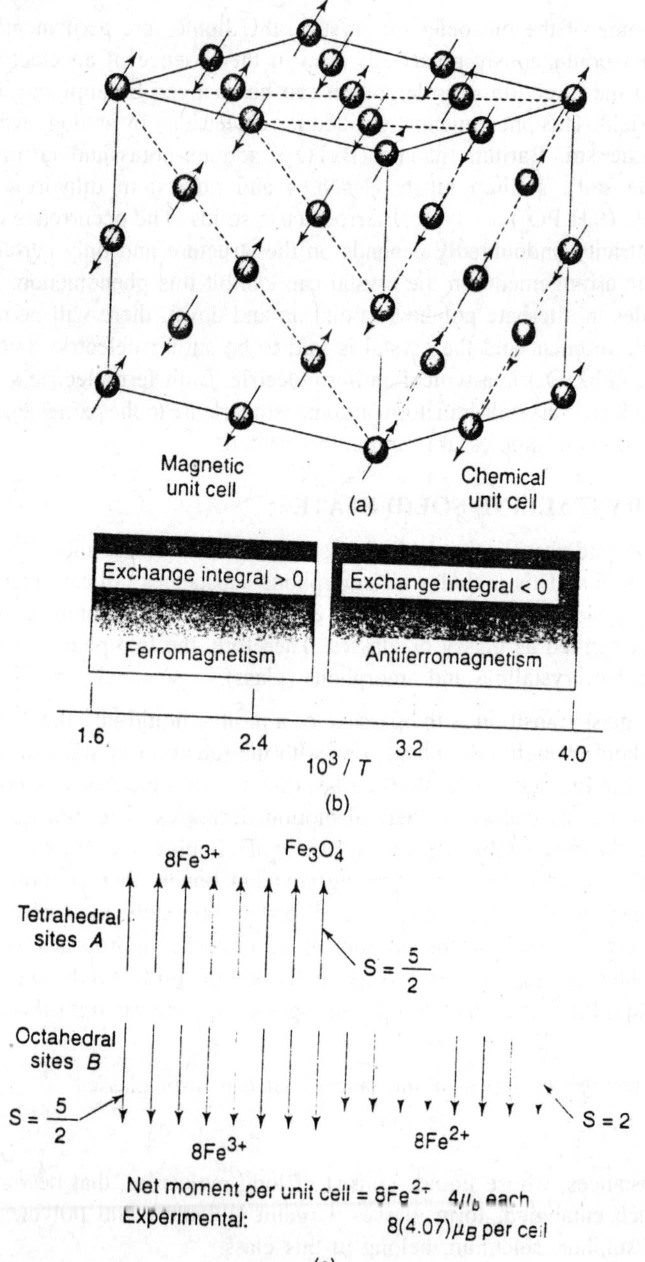
Magnetic
unit cell
Chemical
unit cell
(a)
Exchange integral > 0
Ferromagnetism
Exchange integral < 0
Antiferromagnetism
1.6
2.4
3.2
4.0
10³ / T
(b)
8Fe³⁺
Fe₃O₄
Tetrahedral
sites A
S = 5/2
Octahedral
sites B
S = 5/2
S = 2
8Fe³⁺
8Fe²⁺
Net moment per unit cell = 8Fe²⁺ 4μ_b each
Experimental: 8(4.07)μ_B per cell
(c)

Fig. 2.37

Open Structure Crystals

These substances, which have directed covalent bonds and low coordination number, crystallizes in an open structure. SiO_2, H_2O, oxides of B, Si, Ge, P and As belong to this class. H_2O glasses are formed on extremely rapid cooling or upon condensation at 100°C. Alcohols, glycols and some other hydroxy substance also forms glasses.

In the case of SiO_2, each silicon atom is surrounded tetrahedrally by four oxygen atoms, each of which is bonded to two silicon atoms. The bond length of Si-O is 1.62 A and the ratio of the distance between comers of tetrahedron to the distance from a comer to its centre is 1.64. In the melt Si–O bonds are continually forming and breaking. When the melt is rapidly cooled the particular bond pattern present at that instant is frozen. The resulting glass is still a giant molecule but a rather disordered one. The result is a solid with the rigidity of a crystal but no regular order. In glassy state the tetrahedron order is distorted (Fig. 2.37). The positional disorder existing within each distorted tetrahedron prevents the prediction and assignment of fixed position to the atom.

NON-CRYSTALLINE POLYMERS

Polymers undergo continuous transition from an ideal order to an absolutely disordered state. Very many polymers do not exhibit crystalline state and the crystalline order is also not found in the melt and the solutions of polymers because the thermal motion destroys the crystalline lattice. In the case of rapidly cooling the polymer solution, either a completely amorphous polymer or a body where crystalline and amonaphous region coexist, can be obtained. All the above mentioned cases form a group of non-crystalline bodies eronously called amorphous polymers.

Kargin suggested a definite degree of ordering within the limits of a non-crystalline state. He proposed a bundle model containing a set of macromolecules which are packed almost parallel and have large longitudinal dimensions. Let W(R) be the relative probability W of finding neighbouring atoms at a distance R from a certain fixed atom. For a crystalline substance this function shows constant distances at which the neighbouring atoms are arranged, exhibiting a long range order in the body. For non-crystalline polymers, it completely changes. The first high maximum conforms to the probability of an ordered arrangement of the first particle layer relative to the fixed particle showing a short range order. Such an arrangement results in ordered clusters on associates. Since the height of subsequent maxim is considerably lower, W(R) –1 and hence, the long range order is absent.

In the case of non-crystalline polymers the X-ray patterns give d_1 as 4.5-5Å and d_2 as 10Å. The presence of these maxima show that either macromolecules or segments of macromolecules are packed almost parallel to one another. Therefore, the main feature of non-crystalline polymers is a short-range order up to 10Å and absence of a long-range order.

The employment of electron microscope has shown that in crystalline polymers folding chains form the crystalline lamella, but in amorphous polymers parallel arrangement of chains occur up to 30-100Å, showing a short-range order. There is great deviation from a crystallographic order in the domain. It is assumed that these domains are connected with another by tie chains. Such domain exhibit transition from an amorphous to a crystalline structure. The non-crystalline polymers display a truely amorphous region which are formed by chain ends.

Yet has built a model for amorphous polymers showing domains, the chains, and truely amorphous regions. Kahanon postulated the formation of super domain structure in amorphous polymers (Fig. 2.38).

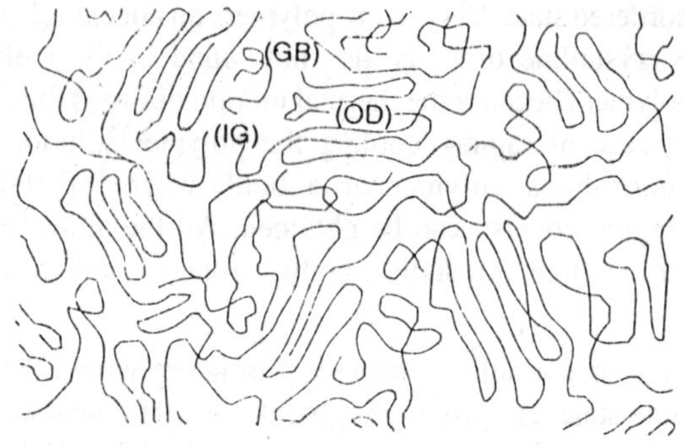

Fig. 2.38

Plastics

Polymers according to their mechanical properties have been classified as elastomers and plastomers. Elastomers are rubbery elastic materials. Plastomers or plastics are macromolecular materials which flow on deformation. The deformation is permanent in plastics but in elastomers the body returns to its initial state instantaneously upon the removal of stress. Normally, in elastomers the chains are very flexible at the room temperature but they are stiff in the case of plastics. This depends on several factors namely, the degree of polymerization, cohesive forces acting between the chains, flexibility of the chains, their branching and cross linking and the degree of crystallinity. The effect of crystalline structure is discussed below.

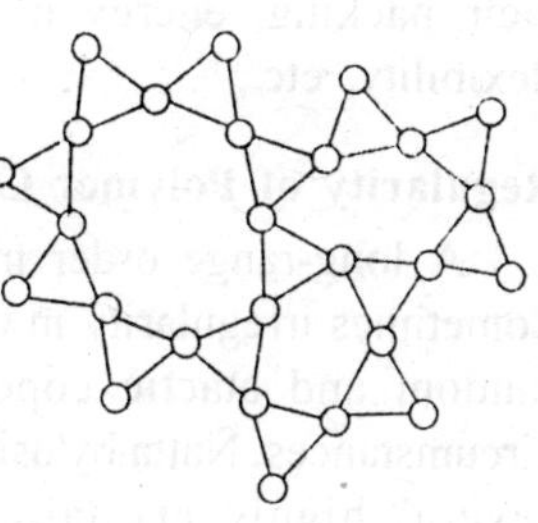

Fig. 2.39

Polymer Crystals

Polymers are products of high molecular mass and their decomposition temperatures are far below their boiling points. Hence, they exist only in liquid or solid state. Although most of the polymers are amorphous, but the crystalline polymers known as isotactic polymers have high melting point and produce fibres of high tensile strength and excellent mechanical properties. A polymer has two types of structural elements viz. monomer units and chains. The long range order is a prerequisite for the crystallization of polymer, not only in the arrangement of units but of chains as well as in three dimensions. This can happen in three ways:

1. Long-range order of arrangement of both chains and monomeric units in three dimensions.
2. Long-range order of arrangement of chains, but the monomeric units are arranged disorderly.
3. Long-range order of the arrangement of units but short range order of chain arrangement.

The high degree of ordering can be achieved in polymers in two ways, *i.e.*, by crystallization or by chain orientation without orientation of monomeric units. These correspond to crystalline and amorphous (glassy) state. It is possible to have a perfect short-range order in chains but the long-range order in arrangement of chains and monomeric units is not very perfect. Therefore, the examples of well ordered liquids and defective crystals are well known.

Crystallization of Polymers

Many polymers are incapable of crystallization under any condition. Their crystallizability is determined by their chemical constitution, *i.e.*, their packing, energy of intermolecular reaction, chain regularity, flexibility, etc.,

Regularity of Polymer Chain

A long-range order must exist in the chain in three dimensions. Sometimes irregularity in the chain may not obstruct crystallization but random and atactic copolymers cannot be crystallized under any circumstances. Natta by using certain catalysts succeeded in synthesizing several highly crystalline vinyl polymers, *e.g.*, polypropylene, polybutylene and polystyrene.

A mixture of the polymers of various degree of orientation was formed and crystalline fraction was separated from amorphous fraction by means of solvents as crystalline polymers are less soluble than amorphous polymers.

Second Order Transition Temperature

A first order transition is one at which a definite change in volume occurs. A second order transition is one at which a change occurs only in the rate of change of volume with temperature. Below T_g, polymers are rigid, glass like material and above T_g they are soft elastometric materials. The second order transitions are caused by thermal agitation releasing portions of the resin chain to rotate around its bonds.

These second order transitions can be measured conveniently by deflection temperature DT tests.

Flexibility of Polymer Chain

Polymers with sufficiently flexible chain can crystallize, but liquid chain polymers crystallize with difficulty because the thermal motion of the momeric, units disturb the established order and the crystal breaks. To crystallize a polymer, it is cooled to a temperature where the motion of its units does not hinder their orientation but excessive cooling may also hinder the rearrangement of the units. Crystallization of each polymer is possible only within a definite temperature interval, *i.e.*, between the glass temperature and flow temperature, *e.g.*, the glass temperature of isotactic polystyrene is about 100°C. Hence, it will only crystallize between 100°-220°C which is its melt temperature.

Packing of Molecules

Polymers obey the principle of close packing and macromolecules must be packed as closely as possible in the crystalline lattice. There are several possibilities for the formation of close packings of polymer chains.

The first crystalline structure is built by close packing of spheres which is exhibited by globular proteins. The formation of such crystalline structures are possible because all the spheres are equal in size, *e.g.*, edestion catalase, iodobenzoyl glycogen.

The second possibility of close packing is the packing of helical macromolecules. The convexities of one helix fit into the concavity of other, *e.g.*, amylose, poly (hexamethylure adipamide), polypeptides.

The third is the packing of long extended chains, *e.g.*, polyamide, polyurethanes, etc. The essential requirement is that the side chain substituents must not hinder regular arrangement of adjacent chains.

Close packing of chain molecules can only be accomplished with flexible chains capable of shifting in parts. Introduction of polar groups gives rise to two opposite effects. On one hand, intramolecular attraction increases, favouring close packing, while on the other hand, the thermodynamic flexibility (potential energy difference between two positions of the chain) of the chain decreases. The chain becomes more rigid. The rate of close packing depends on these two factors. However, with polar groups like –OH or –CN, the attraction effect prevails inspite the fact that the chains are rigid. Since rigid chains closely pack in the extended state, such polymers are highly oriented, *e.g.*, poly (vinylalchol), polyacrylo butyrate. The chains of isotactic polypropylene are highly flexible and form crystalline lattice, but rigid cellulose chains cannot form such lattices under any condition.

Polymer Crystal Cells

The crystalline cell of polymers consists of parts of chains located at definite distances a, b, and c and angles α, β and γ. They do not differ at all from cells formed by low molecular weight compounds and obey all the rules of symmetry.

Polyethylene crystals form orthorhombic unit cells with a = 7.40Å, b = 4.93Å and c = 2.53Å and density equal to 1.0 g/cm^3. All the carbon atoms in the polyethylene cell are arranged on the same plane forming a zigzag plane. In the polyethylene unit cell, chains are arranged along four edges and in the centre of the cell (Fig. 2.40a & b).

In the case of chains having bulky side substituents helical crystals are formed, *e.g.*, polypeptides, polypropylene divinyl and olefin polymers.

One helical loop may contain a different number of monomeric units depending on the nature of a polymer. Natta synthesized 1,2 polybutadiene and similar results were experienced with isotactic polystryrine.

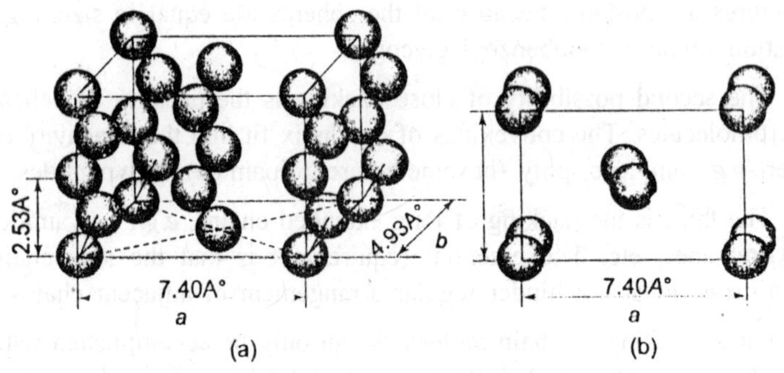

Fig. 2.40

Polymers like sulphur or phosphorus are characterized by polymorphism. Polypropylene may form crystals belonging to monoclinic, hexagonal and triclinic types of symmetry. They are typical first order phase transitions. These transitions between various polymorphous formations take place either upon a change in temperature or under the action of load which brings a change in the parameters of a crystalline cell. As a matter of fact most of the polymers can be crystallized under stretched conditions.

The morphology of a crystalline polymer is determined by the mutual arrangement of the unit cells within the limits of the crystalline state of a body.

Single Crystals

Single or ideal crystals are formed by the parallel transfer of unit cells along edges at distances equal to periods. Fischer first of all ascertained the possibility of high molecular compounds to form monocrystals. Polymer single crystals like other crystals have been found to contain certain defects originating in the packing of chains.

Lamellar Crystals

Polymer crystals are grown by crystallizing from a one per cent solution of polymer under gradual cooling or isothermal aging under the temperature of dilution. The dimensions, shape and regularity of structure of a single crystal depend on the chemical structure of a chain and the physical constraints like temperature, concentration of solution, the cooling rate, the nature of the solvent, etc. The polymer single crystals are monolayer flat plane Lamella which are often rhomboid of 100Å thick and the size of the plate up 1 mp. The axes a and & of a crystalline cell correspond to long and short diagonal of the rhomboid, while axis c along which the polymer chains are directed is perpendicular to the crystal plane. Since the length of the polymer is very big while thickness of the crystal is of the order of about 100Å, the long chain can only fit in the crystal by turning on its surface by 180°. This is known as chain fold conformation. To have an idea in a crystal of 120Å thickness, the fold shall contain 100 carbon atoms. A molecule having a weight of 105 shall fold about 70 times.

The thickness of a lamella, the dimensions of crystallite and degree of defects of the boundary layers of a crystal are affected by the crystallization conditions. Crystals with folded chain conformation can be obtained when super cooling is in the range 15° – 20°C.

Crystallization under high hydrostatic pressure 5-10 × 10^3 atm give extended chain crystals. Polyethylene annealed under a pressure of 700 atm give extended chain crystals. When the melt is cooled under great deformation crystals of different morphological forms can be obtained. Polyhedrons known as hedrites and axialites or avoids are obtained by raising the concentration of the solutions.

Fibrillar Crystals

Fibrillar crystals are obtained by the high rate of evaporation of a solvent from a moderately concentrated solution or at cooling of melt. There are two views about their formation. One section believes that aggregation of rolled lamella is responsible for them while other hold that lamella degenerate in the formation of fibrillar crystals.

In stereospecific polymerization, fibrillar crystals are directly found. The formation is determined by the ratio of the rates of growth of a chain and its crystallization. By changing this ratio crystal either with extended or folded chains are obtained.

Globular Crystals

In globular crystals, macromolecules in coiled conformation form the lattice points. These are found in biopolymers, as a very high degree of uniformity of macromolecules is a pre-requisite for it, *e.g.*, tobacco mosaic virus synthetic polymers do not form them.

Spherulites

Spherulites are obtained when polymers crystallize from melts or concentrated solutions. They are made of crystallites which grow radially from one common centre. Macromolecular chains make at least an angle of 60° with the radius, *i.e.*, they are arranged tangentially to the spherulite radius Spherulites possess anisotropic properties. Their indices of refraction of light are different in radial or tangential directions. The orientation of the crystallographic axes continuously changes along the angular coordinates, A spherulite is called positive when the refractive index in the radial direction is greater than in the tangential one; otherwise it will be a negative spherulite.

In radial Spherulites one of the axis of the crystalline lattice retains a constant direction along all the radial directions. Ring type Spherulites are built of lamellar crystals whose orientation continuously changes along the spherulite radius.

Depending on the degree of cooling, the same polymer may form different types of spherulitis. With low degrees of super cooling ring type Spherulites are obtained while radial Spherulites are obtained by high ones.

The formation of Spherulites passes through several stages. First the crystal nuclei are formed which are scattered throughout the volume of the sample and then independent growth of discrete fibrillar or lamellar crystalline structure grow at the same time in all directions. Then encountering one another they grow into regions filling the entire volumes of the body. Then their boundaries are distorted and they assume the form of a polyhedron.

The structural units in spherulite are bundles of macromolecules made out of chains in an extended conformation. This is possible when several macromolecules try to crystallize simultaneously from both the ends. Such interstructural bonds combine to form separate crystallites inside fibrillar or lamellar crystals.

Oriented State of Polymers

Polymers because of great anisotropy of the dimensions of macromolecules exist in oriented state. In the extreme case all the macromolecules will be oriented parallel to each other, but in reality this can happen on the average as the flexible chains do retain some fold conformations. In the oriented state the difference in amorphous and crystalline polymer vanishes.

Peterlin suggested that in oriented polymers microfibrils of diameter 100-200Å clearly divide crystalline blocks of chains with fold conformation. These chains are divided by defective layers and a large number of intra- and smaller number of interfibrillar chains. The microfibrils continue into larger fibrils to create the macrostructure of oriented bodies of the size 1000Å.

These polymers are important for producing the fibres and films. In the first case an uniaxial orientation takes place while in the second a planar orientation takes place.

Liquid Crystals

Certain substances that contain rod-shaped molecules can exist between the crystalline and liquid state. The forces between these molecules are greatest when they-lie parallel and the crystals of such substances consist of parallel arrays of molecules stacked side by side when the crystal is heated, a point is reached at which the thermal energy is sufficient to break down the crystal lattice but not the parallel arrangement of the molecules. This results into a liquid crystal when the molecules are free to move provided they all remain parallel. At higher temperature the liquid crystal melts to a liquid. Such crystals are called *nematic*.

There are other liquid crystals where the molecules are not only parallel but their ends are aligned. Such crystals are known as *sematic*. In this case the crystals contain layers where the parallel molecules are free to move about in its layer but not in other layers. These crystals form drops in which the surface forms sharp steps with flat risers in between.

The intermolecular attraction in liquid crystals do not act equally in all directions with the result that all the degrees of freedom are not liberated. In a nematic liquid crystal the molecule can rotate only about its long axis and two rotational degrees of freedom are lost. In sematic liquid crystal the molecules are free to move only in the two directions and some translation degree of freedom is lost.

ELASTOMERS

Elastomers are non-crystalline high polymers that have a three dimensional space network structure.

They are built up like a fishing net, the siring of each mesh is replaced by a flexible chains of atoms with only weak forces between separate chains. These chains are prevented from melting by crystalline or by covalent interbonded portions. The distance between the head and tail of the chain is greatest when it is stretched out straight and less when it coils itself. It is therefore, elastic at the room temperature. Unlike steel, stretched elastomer is found to shrink on heating. It is another example of ordered and disordered arrangement.

UNIT CELL OR LATTICE

A unit cell is a convenient block, carved out from a crystal of infinite extension, bound by three pairs of parallel planes. The movement of this block along the three axes reproduces the whole crystal.

Liquids on cooling give out energy, resulting in the decrease of translational energy of molecules, so much so, that they come under the influence of the field of secondary valency forces. The energy at this stage is only sufficient for rotational and vibrational motions of molecules, atoms, or ions, about a position of maximum potential energy, resulting in a solid.

Liquid Crystal Phase Transition

Lennard-Jones *et al.*, developed a model to explain the phase transition between the liquid state and solid crystalline state. They assumed that solids are constituted of two interpenetrating lattices represented by a-sites (+) and β-sites (0) (Fig. 2.41a). In a perfectly ordered state, all the atoms occupy α-sites and holes occupy β-sites. If X is the fraction of atoms on α-sites, the long orders will be defined as 2X-1. With increase in temperature some atoms will migrate to β-sites, which are at a higher potential energy due to repulsive forces that are predominant at distances

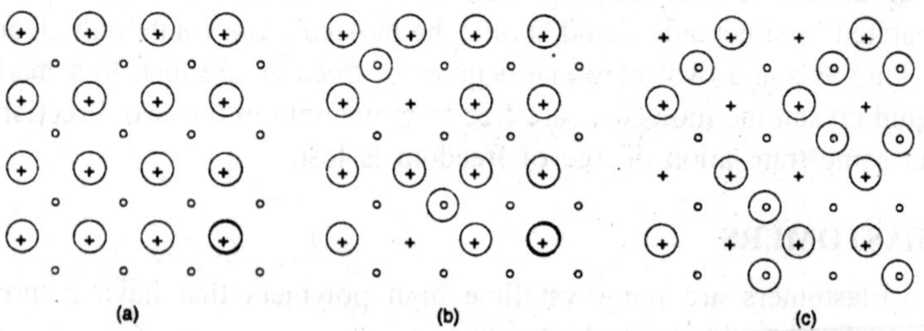

Fig. 2.41

less than the equilibrium distance between α-sites. A few such displacements will still retain long range order (Fig. 2.41 b). At sufficiently high temperatures, the atoms will be randomly distributed between α-sites and β-sites and the long range order will completely disappear (Fig. 2.41c).

The lattice retains a perfect order with all the atoms occupying α-sites even at high pressure over a wide range of temperature.

The lattice can expand without loosing its order provided the kinetic energy of the molecules and the repulsive force between them decreases monotonically with increasing volume at constant temperature.

This increase in volume will create more disorder because of the movement of molecules to β-sites. The initial increase in disorder will be responsible for the increase in pressure due to the increase in repulsive force but when the volume becomes large enough, it would be followed by decrease in pressure as it would permit easy movement of molecules to β-sites.

Thus, this model suggest a phase transition at constant pressure, leading to a critical temperature like gases, above which a liquid-solid transition does not occur. This phenomenon has not yet been observed even at highest pressures attainable.

Crystalline Solid

The molecules, under the influence of cohesive force, will try to orient themselves in an ordered manner in space, provided the rate of cooling is so adjusted that the molecules get enough time to orient and the density of the liquid is not too high.

Thus, a crystalline structure will result. Since there is only a single spatial arrangement of minimum potential energy for all the molecules in a compound, therefore, in the transition from liquid to solid state, the same spatial arrangement is repeated, *i.e.*, we obtain a crystalline lattice of the molecules.

Lattices and Structures

A regular array of points in space is called a lattice.

If identical atoms or molecules occupy fixed position in a lattice, a crystal structure is obtained.

Closed Packed Spheres

The spheres of the same size can be closely packed in one plane as shown in Fig. 2.42a and b. They are so arranged that their centres occupy the comers of an equilateral triangle and each sphere is surrounded by six other spheres forming a regular hexagon, *i.e.*, their coordination number is six. The various depression around each sphere are known as interstitial sites, void spaces or holes.

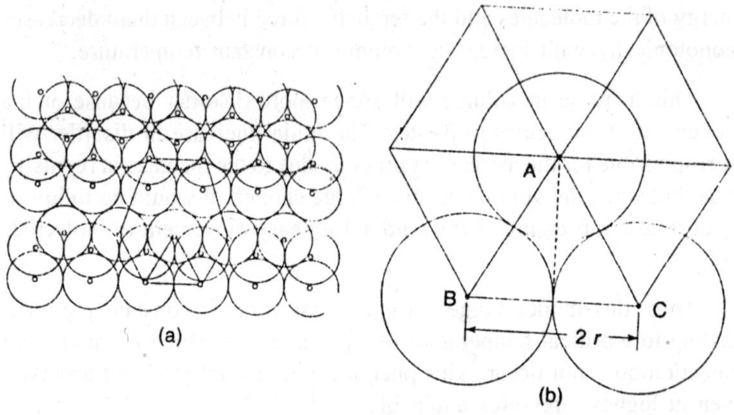

Fig. 2.42

Let us place the second closed packed structure of spheres over the first layer such that small open circles show the position of the centres of the spheres of lower layer and the shaded small circles shown the centres of the spheres of upper layer.

The centres of the upper layer will also occupy the comers of an equilateral triangle. Now the third layer can be placed either at the centre of the first layer or at the depressions of the first layer. The second and the third layer will also be closely packed.

Hexagonal Closest Packing (HCP)

Place the third layer at the centres of first layer and the holes of the second layer. We will obtain a structure which is known as hexagonal closest packed structure (Fig. 2.43).

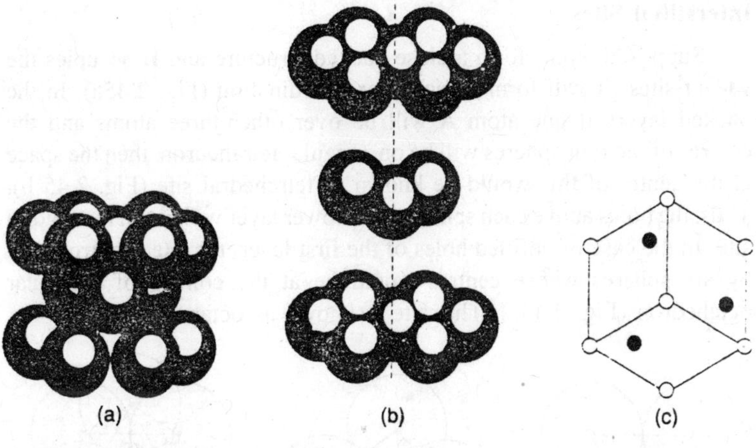

Fig. 2.43

Cubic Closest Packing (CCP)

Place the third layer at the vacant sites of the first layer. Then we will obtain a structure as 123, 123 (Fig. 2.44). This is known as cubic closest packing. Each interstitial site is surrounded by four spheres.

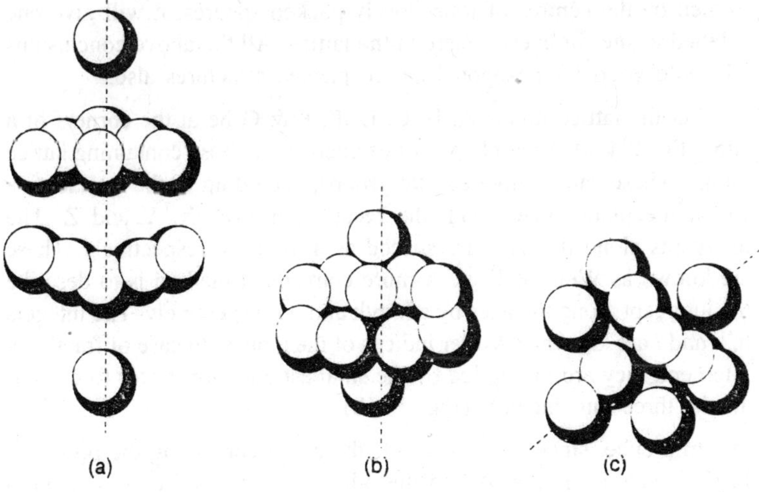

Fig. 2.44

Interstitial Sites

Supposed atoms form a close packed structure and B occupies the vacant sites. It will form a triangular coordination (Fig. 2.45a). In the packed layers if one atom A will be over other three atoms and the centres of the four spheres will be on a regular tetrahedron, then the space at the centre of this would be known as tetrahedral site (Fig. 2.45 b). Thus, the holes above each sphere of the lower layer will give a tetrahedral site. In the case of unfilled holes of the first layer, the site is surrounded by six spheres whose centres would be at the comers of a regular octahedron (Fig. 2.45c). This site is known as octahedral site.

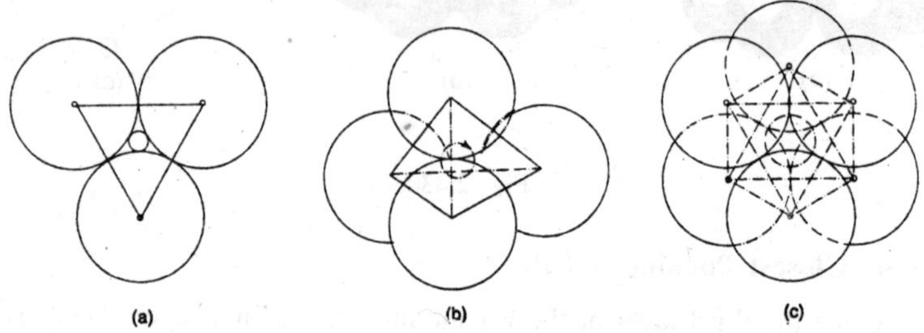

Fig. 2.45

In case the centre of any regular tetrahedron falls on equitorial plane formed by the centres of four closely packed spheres, it will give one octahedral site for every sphere in the lattice. All the above conclusions will hold good for hexagonal closed packed structures also.

Let the lattice points A, B, C, D, E, F & G be at the corners of a cube (Fig. 2.46a).. The only planes of interest are those containing lattice points. These can be divided into groups, according to the intersection which the planes make with the rectangular axes X, Y and Z. The intercepts along the axes are spaced as a, b and c respectively. These are known as Weiss indices. A more convenient method is to describe the intercept along the axes by a/h, b/k and c/*l* respectively. The integers h, k and *l* are known as Miller indices of the planes. In case of fractional intercept, they are multiplied by the smallest common factor to convert all the three into integers (Fig. 2.46b).

In a cube, since a = b = c = 1, the plane containing the points A, F, G, E cuts ox line at A and the other axes at infinity, It is called a (1,00) plane; similarly, the plane containing the points A, B and C is called (1, 1, 1) plane (Fig. 2.46b).

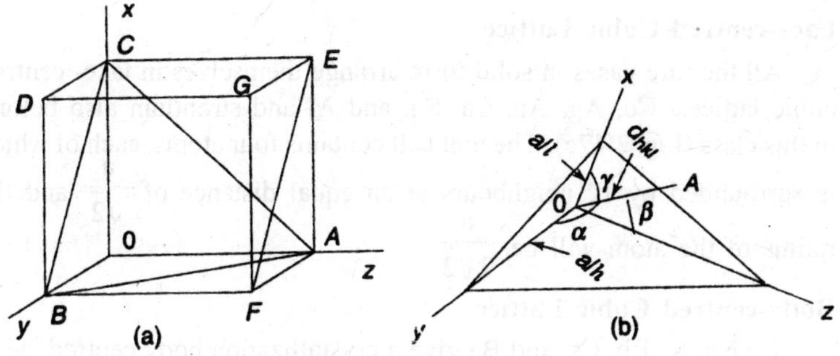

Fig. 2.46

Fig. (2.46b) also gives the angles between the planes. The line OA of length d_{hkl} is the line from the origin perpendicular to the nearest hk*l* plane, a is the angle between OA and the axis X, β is the angle between OA and the axis y, γ is the angle between OA and Z axis. Thus, having known the lengths a, b and c and the angles α, β and γ, a lattice can be fully described.

TYPES OF LATTICE

It can be understood from our knowledge of closest packing that the simplest lattices will be cubic and hexagonal.

Cubic Lattices

There are three types of cubic lattice namely simple cubic lattice, face centre cubic lattice and body centred cubic lattice.

Simple Cubic Lattice

With the exception of the halides of cesium, all alkali halides gives crystals in simple cubic lattice (Fig. 2.47a). In this lattice all the nearest neighbours of any given ion are six ions of opposite charge, each at a distance of 1/2 from the central ion. In case these ions are spheres we find those at the centres of the faces are half outside and those at the comers are only one-eighth inside. Thus, the unit cell will contain $1+\left(12\times\frac{1}{4}\right)=4$ alkali ions and $\left(6\times\frac{1}{2}\right)+\left(8\times\frac{1}{8}\right)$ halide ions.

Fig. 2.47b shows the stacking of submicroscopic cubelets in sodium chloride to produce cubic and octahedral faces.

Face-centred Cubic Lattice

All the rare gases in solid form arrange themselves in face- centred cubic lattices. Cu, Ag, Au, Ca, Sn, and Al and strontium also belong to this class (Fig. 2.47c). The unit cell contains four atoms, each of which is surrounded by 12 neighbours at an equal distance of $\frac{a}{\sqrt{2}}$ and the radius of the atom will be $\frac{a}{2\sqrt{2}}$

Body-centred Cubic Lattice

Li, Na, K, Rb, Cs, and Ba give a crystallization body centred cubic lattice (Fig. 2.47d). The central ion is surrounded by eight equidistant neighbours at a separation of $\sqrt{3a/2}$. The radius r will be $\sqrt{3a/4}$.

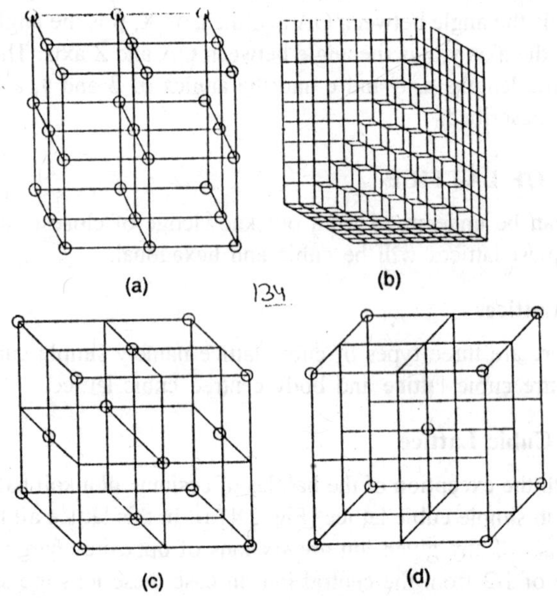

Fig. 2.47

METALS

Non Metals

Non-metallic elements form maximum number of natural and synthetic compounds. We shall try to study the elements in first few series in different groups with the help of ball models.

Allotropes

The existence of an element in more than one form in the same physical state is known as allotropy and the elements are known as allotropes, *e.g.*, o- and p-hydrogen, He-3 and He-4, etc.

Hydrogen

Hydrogen molecule exists in two allotropic forms known as ortho and para hydrogen. Arrange the balls as shown in Fig. 2.48. Let us study the effect of rotation of the balls on their axes.

Fig. 2.48

There are two ways in which the balls can rotate on their axes. A pair of balls will either rotate in the same direction (Fig. 2.48a) or in opposite direction (Fig. 2.48b) in case these balls are assumed as the positive charges (protons) of hydrogen atom. Fig. 2.48a will represent ortho hydrogen and Fig. 2.48b will represent para hydrogen (the electron going around the proton is not shown in the figure).

Hydrogen liquefies at 252.77°C and solidifies at 259.2°C at 55 mm pressure. The crystalline structure of hydrogen depends on the pressure. Under extreme pressure, metallic hydrogen is obtained. Hydrogen gives isotropic crystals belonging to cubic system shapes of either octahedron or hexoctahedron with needle like structure. Table 2.11 gives data about hydrogen crystals by various workers.

Table 2.11

Shape	a_0 Å	C_0 Å
Tetragonal	4.42	3.75
	4.50	3.68
	5.40	5.08
HCP	3.78	6.16
	3.75	6.495

Helium

Helium is an inert gas and has two isotopes in the liquid form viz. He-3 and He-4.

Liquid State : Helium 4 on being cooled below 2.172°K gives a super cooled liquid. The transition to super fluid state is not a first order transition. The phase diagram shows a sharp peak resembling δ. Hence, it is known as δ point (Fig. 2.49).

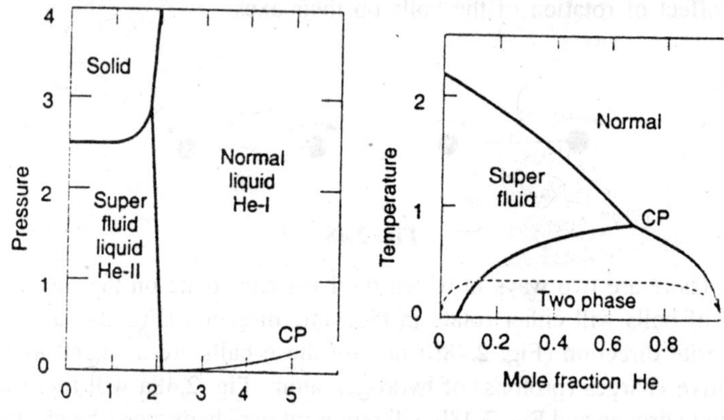

Fig. 2.49

Normal helium 4 liquid is known as helium I and the super fluid as II. He I & He II show a ccp structure at 4.2°K to 4.50°K with 12 atoms at 3.2 A apart. Helium-3, displays three super fluid phases. The phase diagram for liquid and solid mixtures of He-3 and He-4 is given in Fig. 2.49.

Carbon

Carbon is the constituent of numerous naturally occurring substances and also forms the very basis of all organic compounds.

Allotropes of Carbon : Carbon exists in several allotropic modifications, crystalline (diamond Fullerene and graphite) as well as amorphous.

Graphite : Graphite exists as black slippery crystals. Bemal describes the structure of graphite as hexagonal. Infinite layers of atoms of carbon are arranged in the form of hexagons lying in one plane. They are stacked

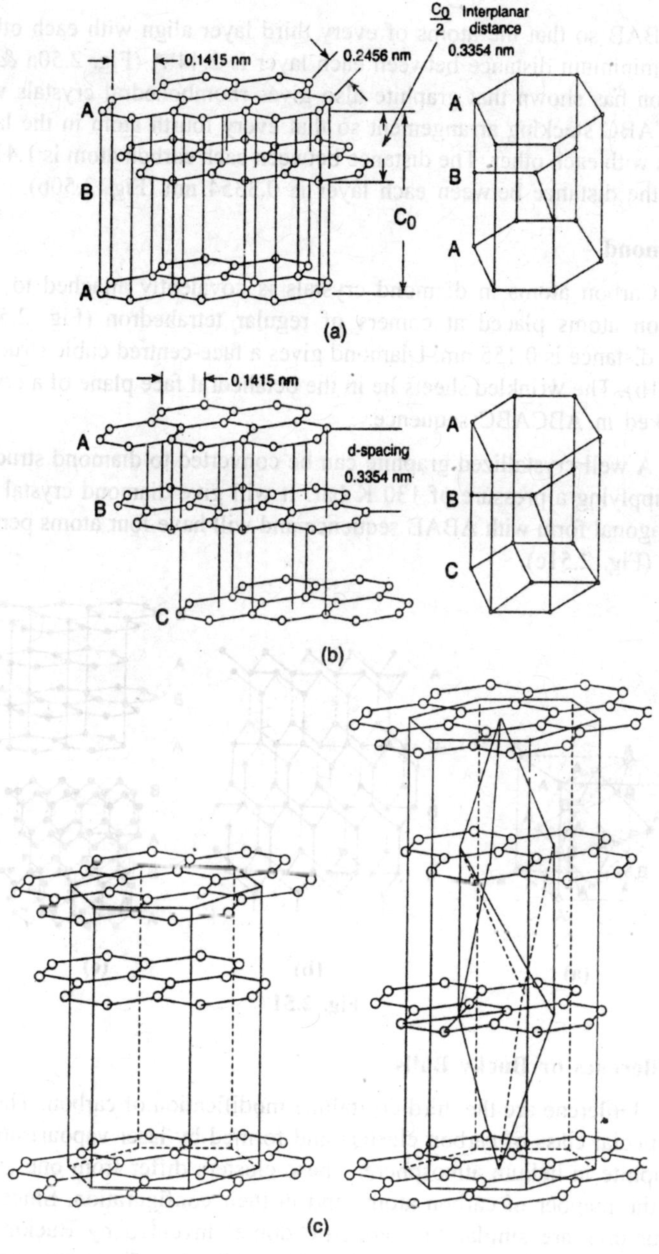
$\frac{C_0}{2}$ Interplanar distance 0.3354 nm
0.1415 nm
0.2456 nm
A
B
A
C_0
A
B
A
(a)
0.1415 nm
A
B
C
d-spacing 0.3354 nm
A
B
C
(b)
(c)

Fig. 2.50

as ABAB so that the atoms of every third layer align with each other. The minimum distance between each layer is 2.44 A (Fig. 2.50a & c). Lipson has shown that graphite also gives rhombohedral crystals with ABCABC stacking arrangement so that every fourth atom in the layer align with each other. The distance between each carbon atom is 1.415Å and the distance between each layer is 0.3354 nm (Fig. 2.50b).

Diamond

Carbon atoms in diamond crystals is covalently attached to four carbon atoms placed at comers of regular tetrahedron (Fig. 2.51a). C-C distance is 0.155 nm. Diamond gives a face-centred cubic structure (2.51b). The wrinkled sheets lie in the octahedral face plane of a crystal stacked in ABCABC sequence.

A well crystallized graphite can be converted to diamond structure by applying a pressure of 130 K bar. It will give diamond crystal in 0 hexagonal form with ABAB sequence and will have four atoms per unit cell (Fig. 2.51c).

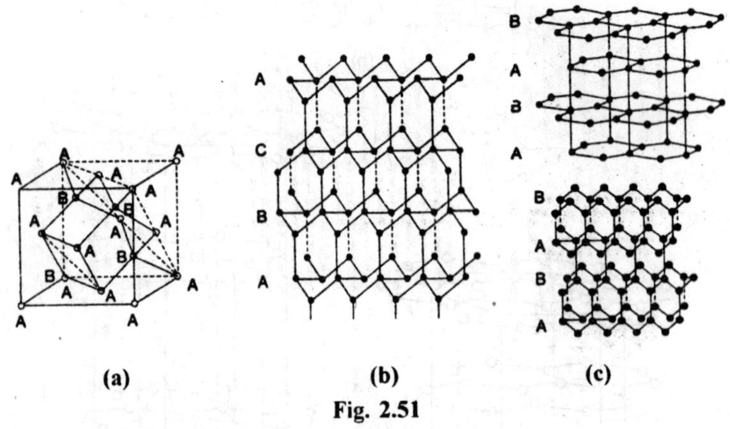

Fig. 2.51

Fullerenes or Bucky Balls

Fullerene are the third crystalline modification of carbon. They are a special class of carbon clusters and formed by laser vapourisation of graphite in helium atmosphere. These clusters differ from one another in the number of carbon atoms and in their configuration. Since these structures are similar to 'geodestic dome' invented by Buckminister Fuller, they were named as buckyball or fullerene. These structures are made up of combination of pentagons and hexagons. They differ in the manner they are combined.

C_{14} and C_{16} are the earlier stable clusters, C_{14} is made up of two fused pentagons giving rise to a octagon which combines with a hexagon, giving octagon type of structure. C_{16} is made up of two pentagons combined with one hexagon, giving a perfect ball structure.

C_{60} finds a unique position. It is made up of 12 pentagons and 20 hexagons, giving a perfect ball like structure. The relative position of carbon atoms in the ball, each and every carbon atom is exactly identical to that of other 59 carbon atoms. This arrangement of carbon atoms imparts highest degree of symmetry in the third dimension resulting in highest stability. Euller has also proved the above mathematically.

Formation

The important question is that how does 60 carbon atoms come together? Evanston of Northern University has suggested the following mechanism:

The carbon atoms first form paired hexagonal rings around which a long chain of carbon atoms extends Fig. 2.52. These paired structures close up in a process of cyclization and form fullerene fragments. This fragment serves as the seed structure along which dangling carbon chain link themselves into pentagons and hexagons. Thus, the chain spirals around the seed structure, forms a spherodial fullerene structure.

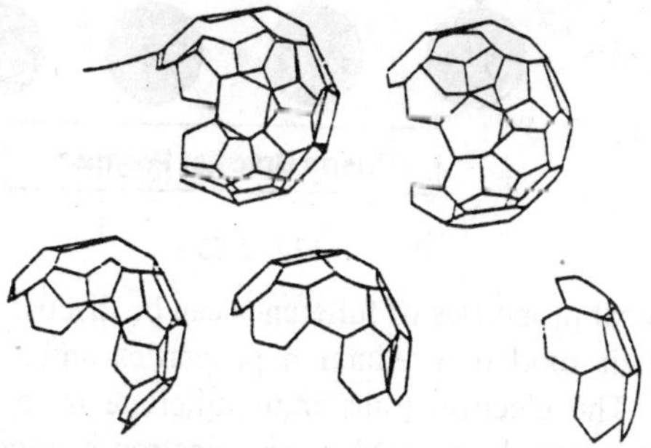

Fig. 2.52 : Smalley's modified 'party line' mechanism for the spontaneous formation of buckminster fullerene

C_{70} fullerene is made up of 12 pentagons and 25 hexagons. They contained a rugby shaped ball Fig. 2.53. The structure of higher fullerenes are given in Fig. 2.54. It is observed that more the structure is away from C_{60}, the more it looses its stability, therefore, C_{70} is less stable than C_{60}.

Properties

C_{60} fullerene is the purest form of carbon. Both diamond and graphite on being cut give surfaces that are instantly covered with hydrogen which ties up the dangling surface bonds:

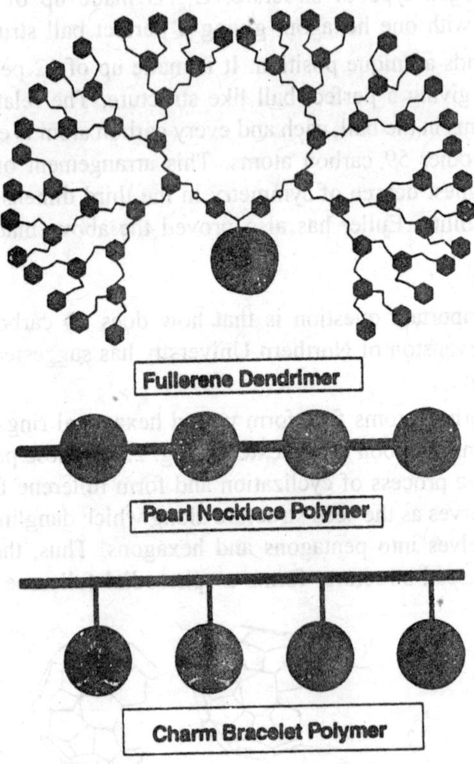

Fig. 2.53

Chemical properties of fullerenes can be discussed in two divisions, namely, their oxidation/reduction properties and its functionalization chemistry. The electron transfer to fullerene is reversible to such an extent that it may be termed as an 'electron sponge'. To explain, C_{60} has a tremendous electron affinity, so that, it can accept upto six electrons, both chemically and electrochemically. Resulting anions, C^{-}_{60}, etc. are remarkably stable and can form radical ion pairs with the donor (D) cation, for example, $[C_{60}]$ [D]. Such ion pairs may exist as stable charge transfer salts. The extra electrons can be easily drawn out from the anion as easily as a sponage is wrung out. C_{60} can be oxidized by super acids

Uke 'magic acid', (FSC_3H: $SbCl_5$) or fluorosulphuric acid (FSO_3H) alone. But the resulting cations are highly unstable because of the inherent electronegativity of it.

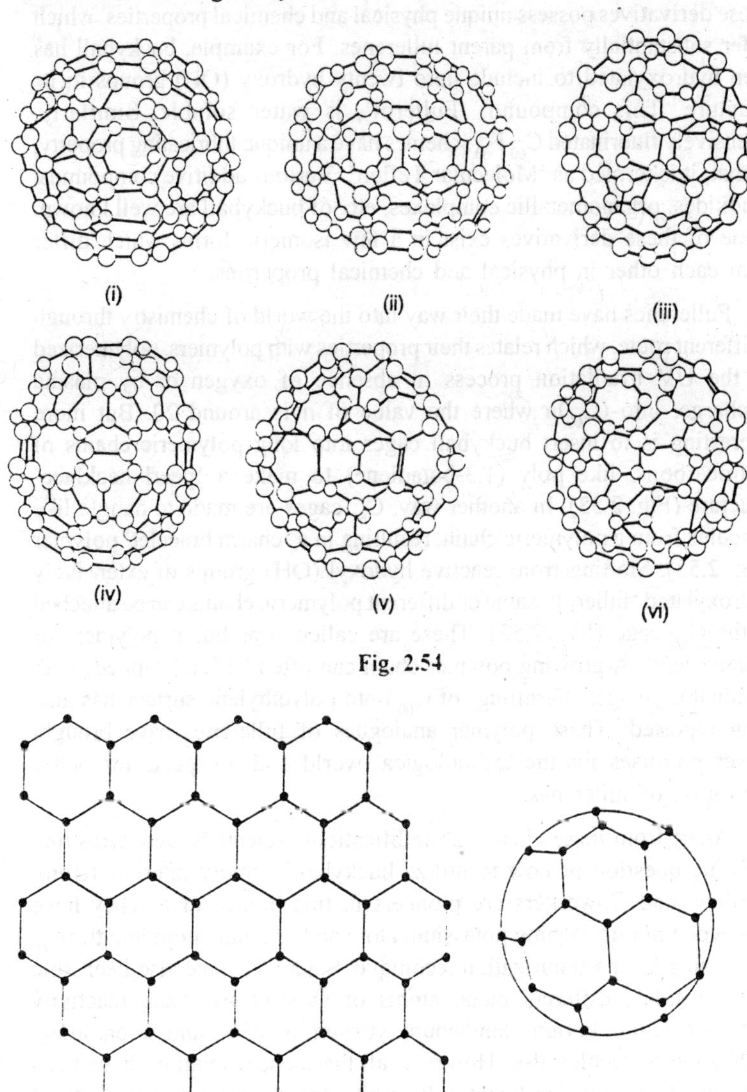

Fig. 2.54

Fig. 2.55

About 12,500 resonating forms are possible for buckyball; still it is less aromatic than benzene. Considerable studies have been done regarding the preparation of derivatives of C_{60} via numerous routes. These derivatives possess unique physical and chemical properties, which differ substantially from parent fullerenes. For example, buckyball has been hydroxylated to include upto twenty hydroxy (OH) groups in its structure. This compound, 'Fullerol', is water soluble. Similarly, extensively fluorinated $C_{60}F_{60}$ schemes have a unique lubricating property, so that, it is termed as 'Molecular Teflon'. Various additive compounds, expoxides, organometallic complexes, etc. of buckyball are well known. Some of these derivatives exist in a few isomeric forms which differ from each other in physical and chemical properties.

Fullerenes have made their way into the world of chemistry through a different route, which relates their properties with polymers, polymerized by the UV irradiation process in absence of oxygen or by plasma discharge, into $(C_{60})n$ where the value of n is around 20. But more interesting is to insert buckyball cages into long polymeric chains of hydrocarbons, like poly (1,3-butadiene) to make a 'pearl necklace' structure (Fig. 2.53). In another way, C_{60} cages are made to dangle like pendants from a polymeric chain, resulting in a 'charm bracelet' polymer (Fig. 2.53). Starting from reactive hydoxyl (OH) groups of extensively hydroxylated 'fullerol', same or different polymeric chains can be attached to the C_{60} cage (Fig. 2.53). These are called 'star burst' polymers or 'dendrimers'. A growing polymer chain can often be 'end capped' with terminal C_{60} cages. 'Grafting' of C_{60} onto polyethylene surface has also been reported. These polymer analogues of fullerenes have brought newer promises for the technological world and prospects for better utilization of fullerenes.

Apart from these chemical modifications, scientists were wrestling with the question of how to utilize buckyball's empty interior. Kroto, Smalley, and Coworkers are pioneers in this branch also. They have succeeded in imprisoning potassium atom and caesium atom into the C_{60} cage using laser vapourization technique. Scientists have also been able to encapsulate different metal atoms of varying sizes and reactivity namely calcium, barium, lanthanum, yttrium, uranium and so on, along with some molecules also. However, all these encapsulations have been performed without significantly disturbing the structure and property of the fullerene molecules. Properties of these entrapped molecules while 'residing' inside the buckyball cage (Fig. 2.56) are opening the door to a new 'nanoworid'. ('Nano' is a word used to describe matter at the scale of 10^{-9} m or one thousand millionth of a meter.). Whether a fullerene cavity can serve as the smallest vessel in the world to trap a reactive atom in its atomic state is also a very interesting subject for scientists.

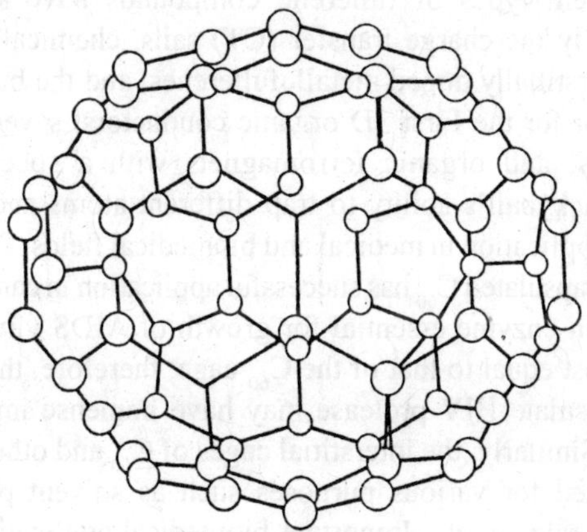

Fig. 2.56

In recent days scientists are getting interested in the novel electrical and electronic properties of fullerenes. In different compounds, fullerene changes its property from being an insulator to turn into a superconductor! To explain, buckyball itself is an insulator in its pure form (band gap = 2.2 ev). But, it becomes a semiconductor when suitably 'doped'. This doping may involve putting a dopant atom or molecule inside the cage or some other appropriate route forming 'buckide' salts. 'Fulleride' or 'buckide' salts conduct electricity; conductivity increase with increased metal doping and reaches a maxima at a C_{60} metal ratio of 1:3. With further increasing metal doping the buckide salts start to lose conductivity, $M_6 C_{60}$ (M is a metal atom) being an insulator. What is more important is that $M_3 C_{60}$ is a high temperature superconductor, with a Tc of 18K to 28K. Mixed doping with Cs and Rb results in the formation of a superconductor $Cs_2 RbC_{60}$, with a Tc of 33K. However, all these buckide salts are three dimensional superconductors as compared to previously known two dimensional cuprate superconductors. Discovery of superconductivity in buckides has given a new impetus to fullerene research. Apart from being 'doped', buckyball itself can dope organic conductors/semiconductors. For example, organic polymers like polyaniline, polyacetylene, etc., can be doped with buckyball to induce semiconductivity in them. These polymers, when otherwise doped, behave as 2D semiconductor. In India, DR C.N.R. Rao of the Jawaharalal Nehru Centre for Advanced Research, Bangalore and his group is also working with superconductivity in fullerides.

Four different types of fullerene compounds have so far been discussed, namely the charge transfer (CT) salts, chemically modified derivatives, interstitially doped metallofullerenes, and the buckide salts. CT salts account for the First 3D organic conductors, several high, Te superconductors, and organic ferromagnets with a specified Curie temperature. Buckyball's ability to trap different atoms and molecules has successful application in medical and biomedical fields. For example, radioisotope encapsulated C_{60} has successful application in cancer therapy. HIV protease, an enzyme essential for growth of AIDS virus HIV, has a diameter almost equal to that of the C_{60} cage; therefore, the capability of C_{60} to encapsulatc HIV protease may have immense importance in AIDS therapy. Similarly, the interstitial cages of C_{60} and other fullerenes are being utilized for various purposes such as solvent preservation, controlled drug release, etc. Important biomedical application of water soluble fullerols has also been suggested.

Important applications of catalytic amounts of buckyball have been found. It can improve antiwear, antiseize, antifriction properties of lubricating oils. In petrochemical refining industries, efficient production of many hydrocarbon products are often hindered by undesirable polymerization of reactive olefins. Fullerene, in ppm (parts per million) level, can inhibit such thermal polymerizations and thereby improve the refining efficiency. Dissolved Fullerene inhibits coke forming on a catalyst and may therefore, be used as an anticoking agent. Preparation of glassy thin Films containing C_{60} could offer the possibility of producing filters against optical damage.

The boundary between theory on a catalyst and practise is becoming thinner as the world of fullerenes is being explored. However, all these promises may die within the four walls of laboratory, until and unless the buckyball can be synthesised in plenty. In view of Dr Kroto, C_{60} should exist in abundance in interstellar space, arising from chemical reactions in the outer atmospheres of carbon rich red giant stars. The prediction has been evaluated by spectroscopic observations of astrophysicist Webstar of Edinburgh Royal Observatory. Besides, trace amounts of C_{60} have been found in coal deposits and in some geological stone samples.

In 1966, Dareid F. Jones, published in a paper 'New Scientist' suggesting that the fullerene C_{60} is made up of curbed graphite sheet. But Leonard Euller has proved that it is not possible. Sumito ijima has suggested that curbed graphite sheet will give a cylindrical tube.

These are hollow, three dimensional layered structures called nanotubes. These are of the order of ten to a thousand or more nanometers in length, a few nanometers in diameter and posses several tens of such layer. One side of it is closed with fullerene type of hemispheres. They are stiffer than any other known materials. They serve as conducting nanowires when stuffed with molten conductor. It is expected to use them for chemical and molecular reactions inside these test tubes. In the synthesis of nanotubes, hollow polyhedral particles of diameters 5.50 mm are produced. Motallow fullerene like lanthanum carbide (LaC) are produced. The encapsulation into nanoparticles or nanotubes may prove to be a useful method for protecting sensitive materials.

Fullerene and nanoparticles form 'Giant Fullerene'. These have a diameter of 1.2 mm. Thus, these have opened a new and very wide field of research.

Amorphous Carbon

Amorphous carbon exists in many forms like coal, lignite, gelsonite, coke, gas carbon, charcoal (animal, wood, sugar), lamp black and soot. All these are characterized by imperfect graphite structure. The parallel layer planes are not overlaid to their perpendicular axis, but the angular displacement of one layer with respect to the other is random and the layers overlap one another irregularly. This is known as turbostratic structure. The distance of separation is of the order of 0.35 to 0.365 nm and C-C distance is 0.142 nmboth in graphite and amorphous carbon.

Nitrogen

Nitrogen is an inert gas and does not form polymers in the elemental form. Ammonium cyanide NH_4CN at –125°C forms isochains.

Nitrogen liquefies at –195.8°C and solidifies at –209°C. In solid form nitrogen exists in two allotropic forms namely a-nitrogen and α-nitrogen.

α-nitrogen gives F.C.C. crystals at 4.2°K with a = 5.644 A and Z= 4.

β-nitrogen give hexagonal crystals at –209.9°C with a = 4.039 A and C = 6.67 A. N-N distance is 1.067 A.

Oxygen

Oxygen is the most reactive element and exists in two allotropic forms namely O_2 and O_3 (ozone).

Oxygen liquefies at –182.5°C. Dewar obtaineda hard pale blue solid by cooling liquid oxygen to –227°C at 0.9 mm pressure. In the solid form oxygen gives three allotropic forms namely α–, β–, γ oxygen.

α-oxygen-249.5°C

β-oxygen-230.5°C

γ-oxygen.

Oxygen crystallizes in hexagonal shape when crystallized in boiling hydrogen.

Ozone boils at –111.9°C and solidifies at –192.7°C, but the structure of its crystals is not well established. A modification of ozone consists of oxygen isochains.

Halogens

Of the four elements off halogen family found in nature, fluorine and chlorine are gases, bromine is a liquid and iodine is solid.

Halogens in the solid form given othorhombic crystals and all atoms exist in tetramolecular units (Fig. 2.57).

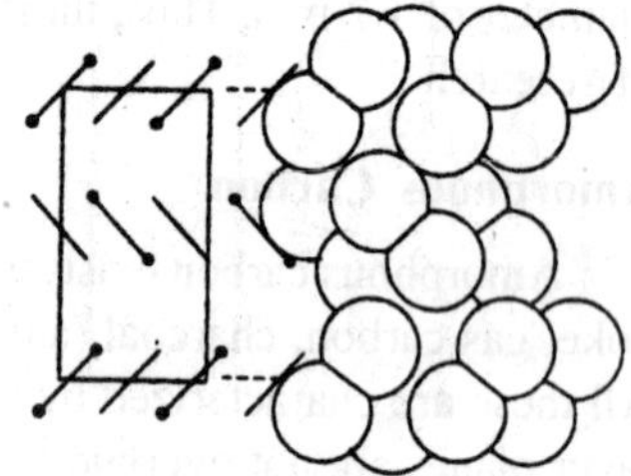

Fig. 2.57

Fluorine exists as a yellowish orange liquid at –188°C, yellowish solid at 220°C and white solid at –228°C. It given mono clinic crystals at –233°C.

The chlorine and other halogen atoms form layers parallel to 0/0 axis. In I_2 the molecules are grouped in layers parallel to ac plane. There are 16 chlorine atoms in a tetragonal unit. Each chlorine atom has two neighbours at a distance of 3.34Å. The distance between Cl-Cl layers is 3.69Å. The atomic separation between bromine molecules is 3.30Å. Rapid evaporation of rhombic bromine crystals give monoclinic crystals.

Silicon

Silicon is inert at room temperature. At the atmospheric pressure, its crystal lattice is the same as that of diamond, *i.e.*, two interpenetrated face-centred cubes. At high pressures but less than 15000 aim., the FCC structure is converted to BCC structure. Vapours of silicon at 500°C on cooling give amorphous silicon. The crystals grown from vapours give octahedral structure. Silicon melts at 1685ÅK and boils at 3900ÅK.

The elemental silicon gives polymer in the solid state. However, silanes $H(SiH_2)_nH$ give isochains with n equal to 45. $(Si\ H)_n$ exists as high molecular mass polymer with six-membered flat rings probably as phyllo-poly (Silanes).

Phosphorus

Phosphorus in the solid state exists in three allotropic forms namely white phosphorus, red phosphorus, and black phosphorus. The solid obtained by condensing the vapour is white phosphorus which gives cubic crystals with a = 2.21Å and is known as α-phosphorus. β-phosphorus is hexagonal tetrahedral in structure composed of four atoms. Each atom occupies the comer of a regular tetrahedron (Fig. 2.58a).

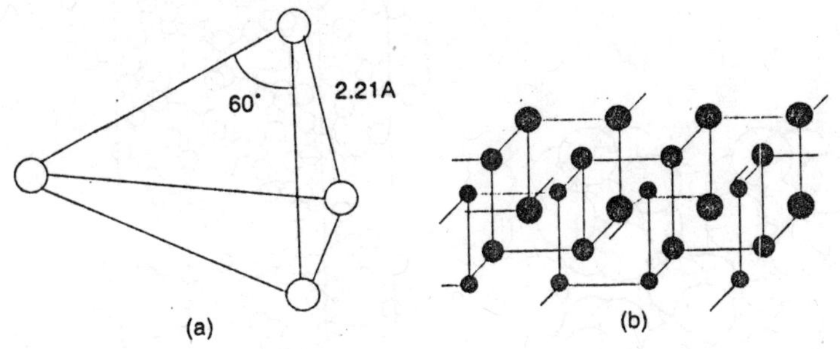

Fig. 2.58

White phosphorus upon heating or irradiation is converted to red phosphorus. Its structure is not well established.

Black phosphorus is obtained by careful crystallization of white-phosphorus or subjecting it to high pressures. It gives polymeric isochain of phosphorus atoms (Fig. 2.58b). The orthorhombic phosphorus has a, b & c equal to 3.313Å, 10.478Å, 4.376Å respectively and Z as 8.

Sulphur

Sulphur occurs in a number of allotropic modifications of formulae $(S\text{-}S)_n$ where n can be as great as 10. Sulphur atoms are linked by covalent bonds having bond angle S-S-S as 105°. Sulphur forms isochains in the melt. Sulphur cations form S_n^{2+} (with n = 4, 8, 16) and sulphur anions form S_n^{2-} (with n = 4,9).

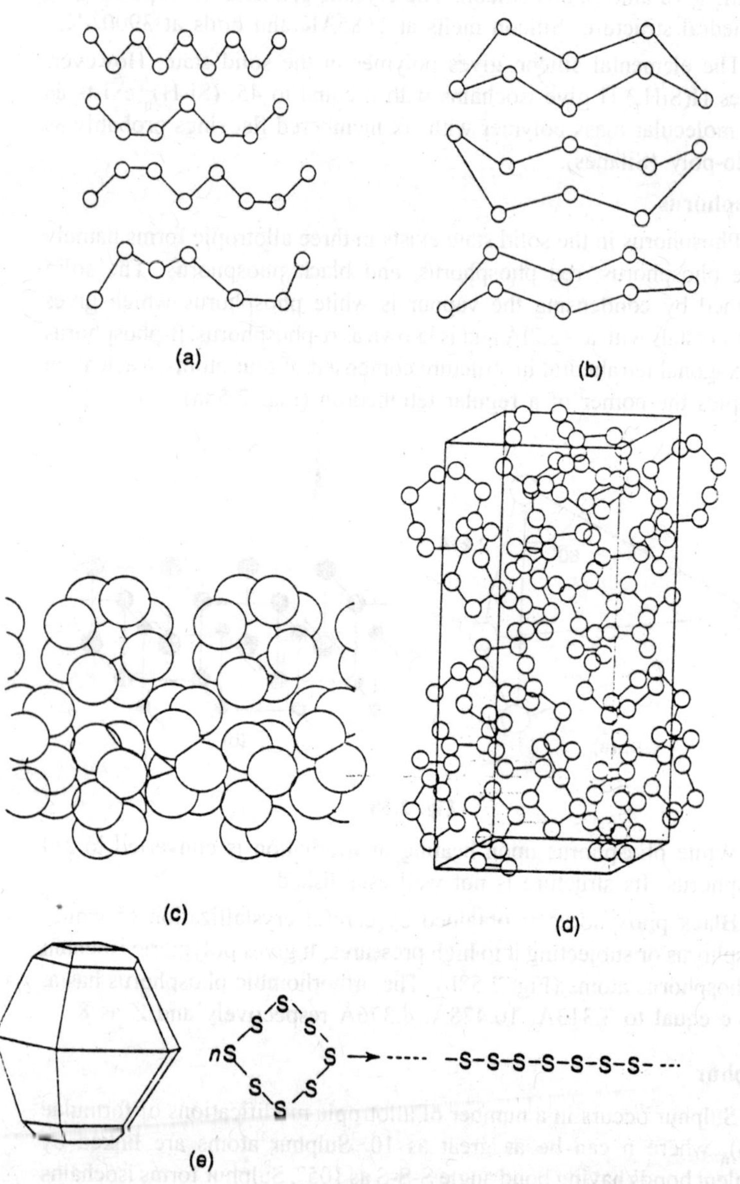

Fig. 2.59

Sulphur in the solid form gives Sδ, Sμ, and Sn. Of these Sπ is stable. In molten sulphur, on solidiftcation, S changes to Sμ. The molecular structure of Sπ is an octatomic sulphur chain (Fig. 2.59a). Sλ, gives a long polymerized chain of sulphur atoms (Fig. 2.59b). 5A gives a crown shaped octatomic sulphur ring S_{R_8} (Fig. 2.59c).

Sulphur exists in two different types of crystals namely α-sulphur (Fig. 2.59d) and β-sulphur. α-sulphur gives rhombic or octahedral crystals, with a as 10.83A and c as 4.26A. This is stable up to 95.5°C. β-sulphur gives monoclinic or prismatic crystals with a as 8.57A, b as 13.05Å. δ-Sulphur is amorphous in nature. Other forms of sulphur are hexatomic sulphur and modifications of catena-poly sulphur.

In liquid state, the chains of sulphur are broken at higher temperature. When the molten mass is poured in cold water, it gives plastic sulphur. Sulphur above 250°C on chilling gives thin rods which can be stretched 20 times the length at the room temperature. The resulting fibre Sϕ consists of Sγ(S_8 rings) and S4 consists of long chain molecules.

IONIC CRYSTALS

We have been dealing throughout with the structures having the balls of the same size. The majority of compounds consist of different elements. We shall now try to build up structures consisting of balls of different sizes. The simplest will be those structures which have balls of two different sizes.

Radius Ratio

Radius ratios of the two different balls play an important role in deciding the shape of the ultimate structure. It is useful to consider that the smaller ball occupies holes in a closed packed lattice of the larger balls. A closed packed lattice contains triangular holes (Z = 3), tetrahedral holes (Z = 4) and octahedral holes (Z = 6) (Fig. 2.60). Consideration of the geometry of the holes leads to the observation that for given larger spheres of unit radius, small spheres that can fill the holes should have the radius.

Crystals

The simplest structures are of AX type, where A represents a ball of larger diameter and X represents a ball of smaller diameter. Balls with two different diameters can be arranged to give structures like AX, AX_2 and A_mX_n. Some of the important structures. The critical ratios are:

Cubic (8) 0.732 $\rightleftharpoons$ Octahedral (6) 0.414 $\rightleftharpoons$ Tetrahedral (4) 0.225 $\rightleftharpoons$ Tringular (3) 0.155 $\rightleftharpoons$ Linear (2).

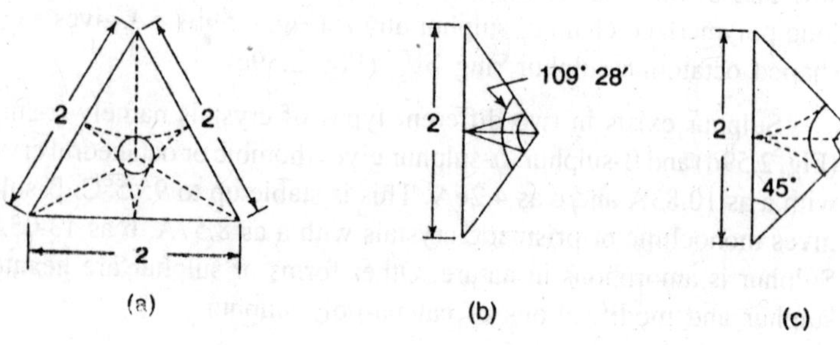

Fig. 2.60 (a)

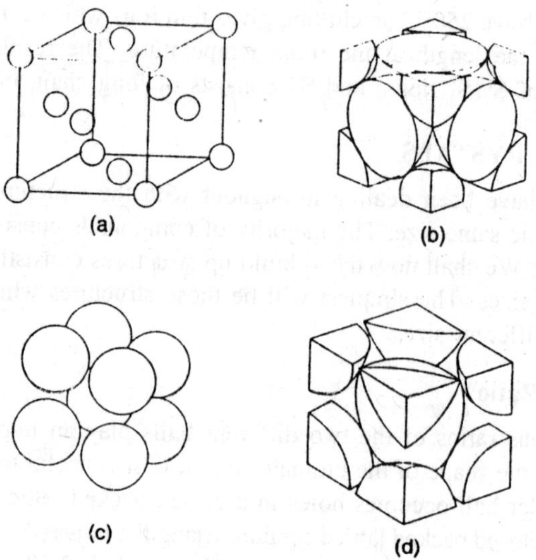

Fig. 2.60 (b) : (a, b, c, d) Different types of structures

(a) A face centred cubic structure or cubic close-packed structure.

(b) Three-dimensional view showing four atoms per unit cell in case of cubic close-packed structure.

(c) Monoatomic body-centred cubic structure.

(d) Three-dimensional view showing two atoms per unit cell in α-body-centred cubic structure.

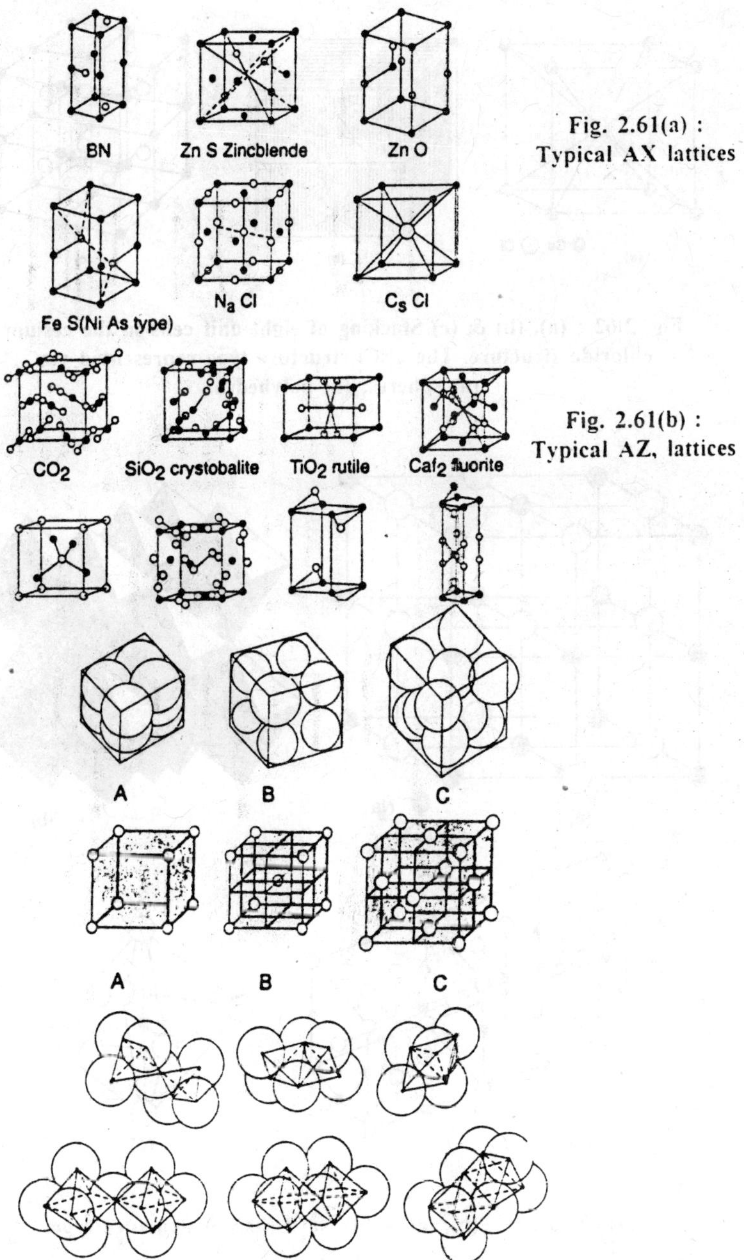

Fig. 2.61(a) : Typical AX lattices

Fig. 2.61(b) : Typical AZ, lattices

Fig. 2.61(c) : The sharing of a corner, an edge, and a ace by a pair of tetrahedra and by a pair of octahedra.

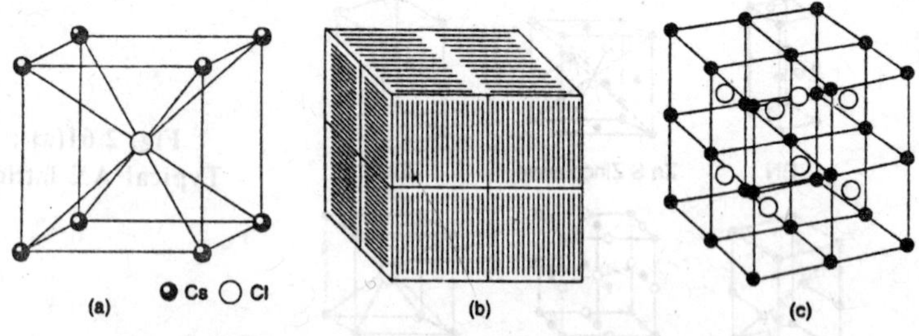

Fig. 2.62 : (a), (b) & (c) Stacking of eight unit cells in the cesium chloride structure. The CsCl structure type represented viz. (a) spheres; (b) polyhedra.

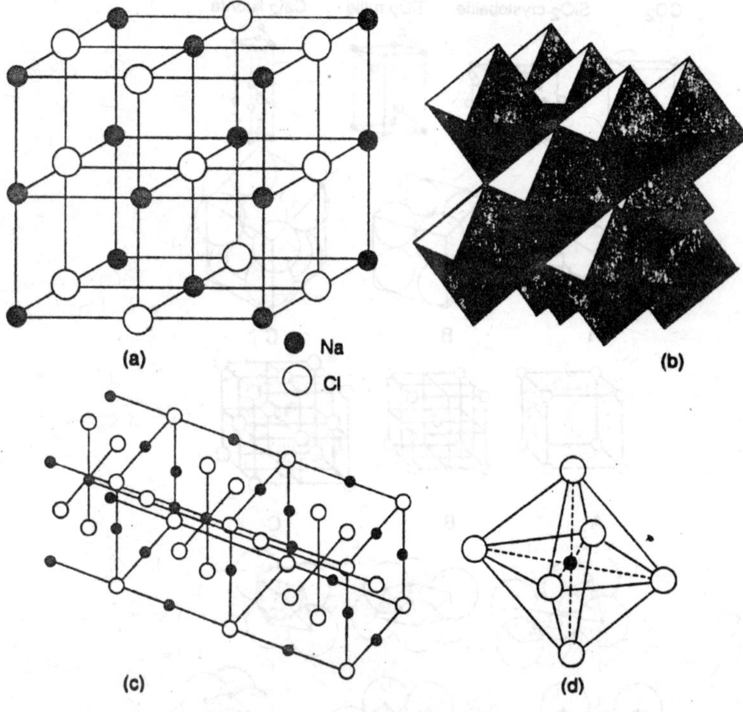

Fig. 2.63 : (a), (b), (c) & (d) NaCl (halite) structure type.

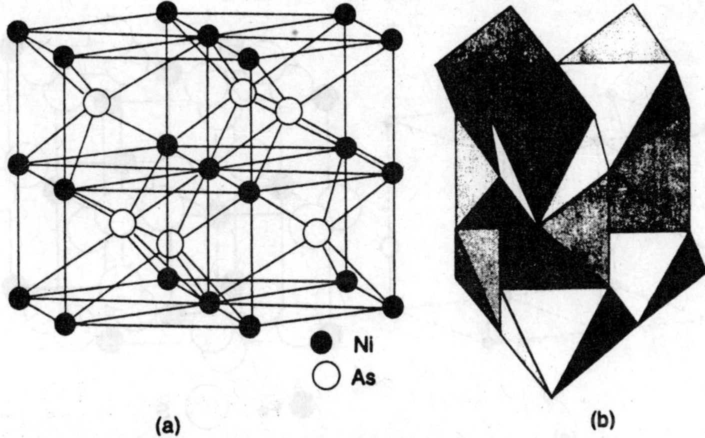

Fig. 2.64 : (a), (b) The niAs (nicolite) structure type.

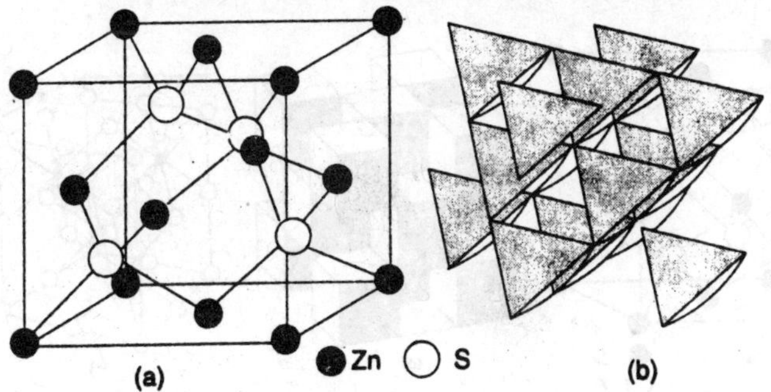

Fig. 2.65 : The ZnS (sphaleriate) structure type.

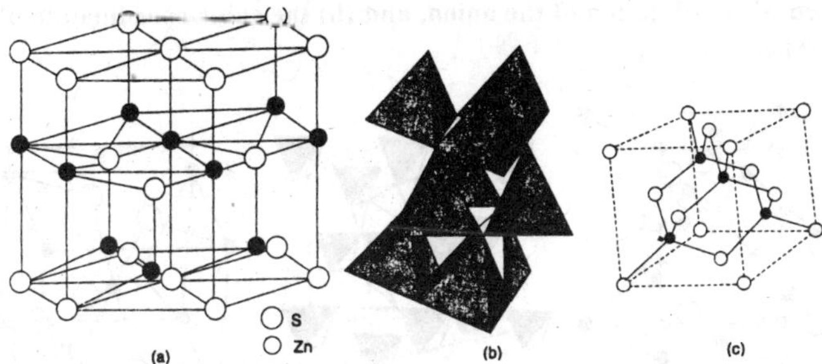

Fig. 2.66 : (a), (b) Wuartzite ZnS structure type; (c) Tetrahedral coordination in the zincblenda (a) and wuartzite (b) Forms of zinc sulfide. Note the similarity of these structures with those of diamond and ice.

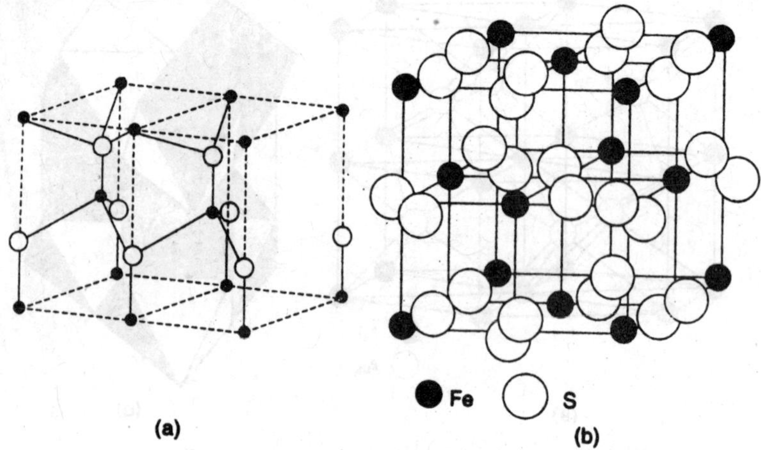

Fig. 2.67 : (a), (b) The FeS_2 (pyrite) structure type.

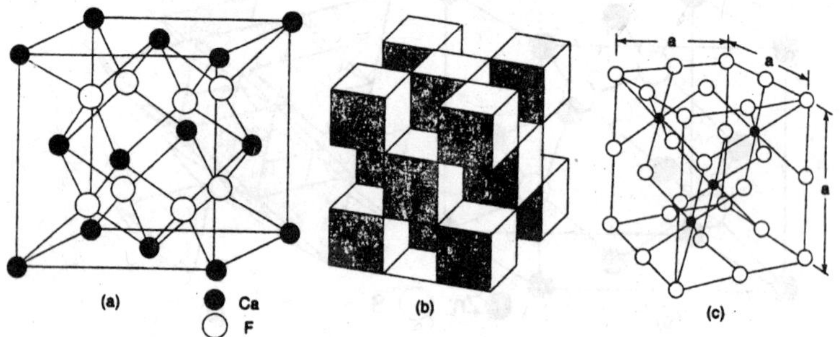

Fig. 2.68 : (a), (b) & (c) Alternative unit cells of flourite to show (a) the tetrahedral coordination of the anion, and (b) the cubic coordination of the CaF_2 type.

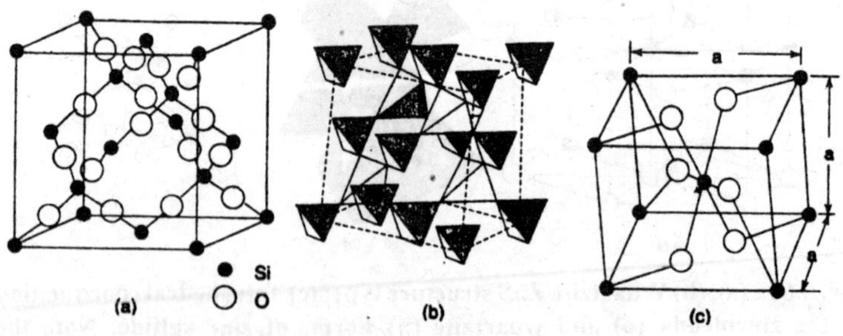

Fig. 2.69 : (a), (b) & (c) SiO_2 type.

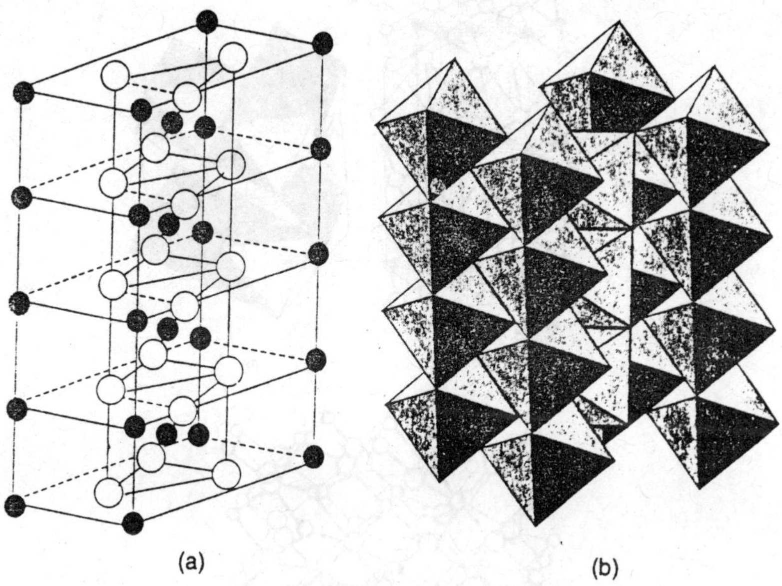

Fig. 2.70 : (a) & (b) The rutile TiO_2 structure type.

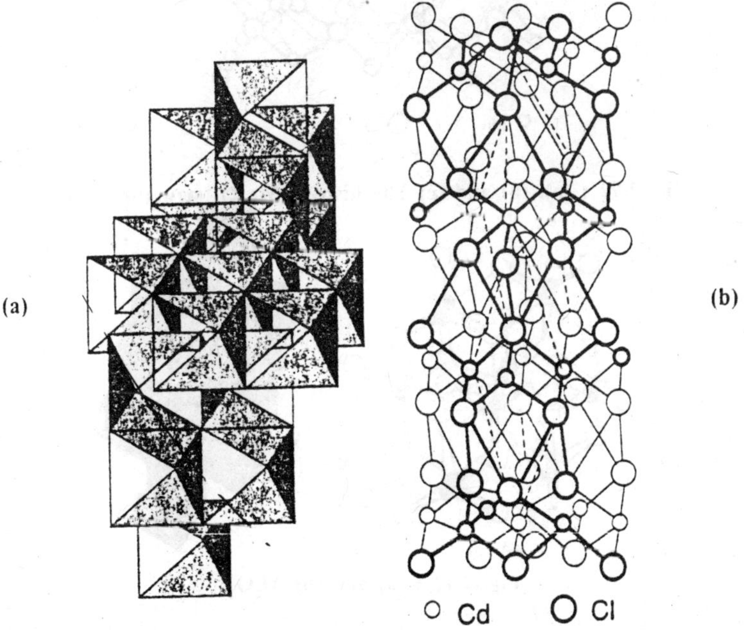

Fig. 2.71 : (a) & (b) The cadmium chloride $CdCl_2$ structure type.

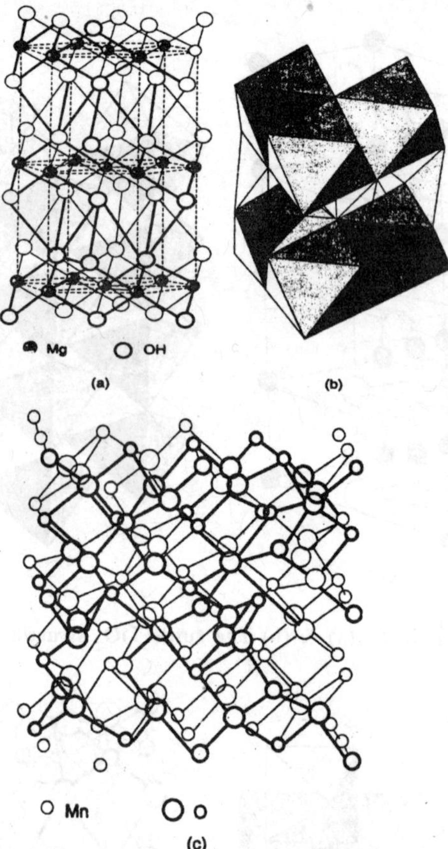

Fig. 2.72 : (a), (b) and (c) The hisbyin Mn_2O_3 structure type.

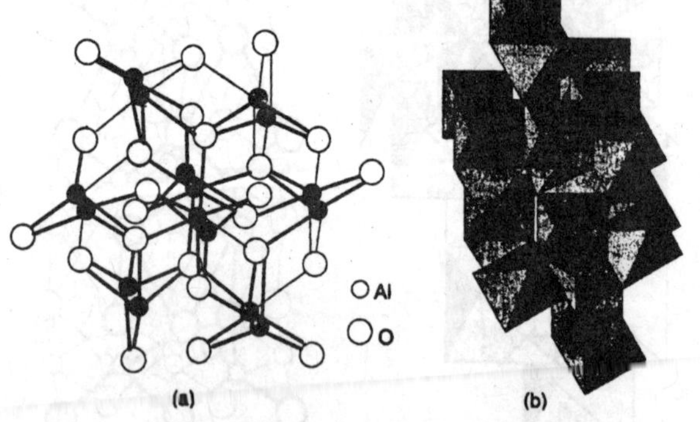

Fig. 2.73 : (a) and (b) Al_2O_3.

TYPES OF PLASTIC STATE

The liquids on cooling, try to attain a minimum potential energy and thus, form crystal. If it so happens that after some time the process of crystallization stops and super cooling starts, we get a series of crystalline structures included in the super cooled structures. Such a state is known as plastic state. Thus, polymers whose chains are stiff at room temperature are called plastic.

Macromolecules

Macromolecules are molecules built from a large number of atoms. They can be of natural origin like cellulose, proteins, etc., or they may be produced synthetically like poly (ethylene) or silicones. All macromolecules consist of at least one chain of atoms bonded together, not necessarily by covalent bond, but sometimes by coordinate bonds or electron deficient bonds.

This is known 'as main-chain which consists of constitutional units and of at least two end groups. A polymer is a substance composed of molecules characterized by the multiple repetition of one or more species of atoms linked to each other in amounts sufficient to provide a set of properties that do not vary markedly with the addition or removal of one or a few constitutional units (IUPAC).

Constitutional Units

A constitutional unit is the smallest unit whose repetition completely describes the main chain structure. Hence, it is known as the constitutional repeating unit.

Monomeric Unit

The monomeric unit (or mer. or base unit) is the largest constitutional unit contributed by a monomer during a polymerization process.

End Groups

End groups are the groups that occur at the end of a macromolecular chain. Linear chains have two end groups but the star shaped polymer with four arms has four end groups. The end groups of a polymer yield information on the mechanism of synthesis and this information may be used to determine the molar mass or the degree of branching.

Polymer Homologous Series

If the molecules have identical constitutional repeating units but differ only in the number of such units per macromolecule, the macromolecules are known as belonging to polymer homologous series. The number of monomeric units joined together in a macromolecular chain is known as the degree of polymerization of a molecule. In exceptional cases the degree of polymerization is identical with the chain link number, which gives the number of atoms joined sequentially together in a chain.

Polymers with a small number of constitutional units and unspecified end groups are known as *oligomers*. Small refers to 2-20 constitutional units in synthetic polymers and several hundred units in nucleotide chemistry. Telomers are oligomers with end groups consisting of fragments of chain transfer agents. Telchelic polymers are oligomers with known functional end groups.

DEFECTS OR DEFORMATIONS

In actual practice it is not possible to arrange all the balls in positions required by specific structure. These balls slip past each other to produce what are known as defects. These defects may be point defects or linear deformations.

Points of Deformations

The defects may not alter the stoichiometry of the crystal, but the balls may move within the crystal itself. Normally these balls will move

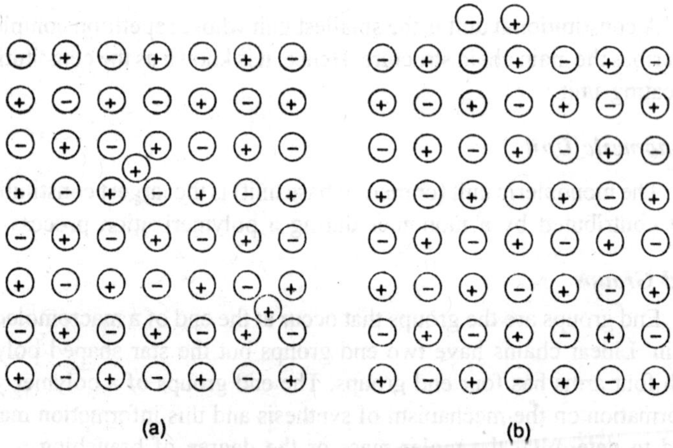

(a) (b)

Fig. 2.74

from an occupied position into a vacant position. These defects are called intrinsic defects. These defects are of two types. In one, the ball moves from its lattice position to an interstitial position. This type of defect consists of the vacancy plus the interstitial and is known as Frenkel defect (Fig. 2.74a).

In the second defect a pair of balls of two different sizes leave the position in the crystal and occupy different sites. This type of defect is known as Schottky defect (Fig. 2.74b).

Linear Defects: Dislocations

The shifting of the balls from atomic sites in the crystal without separation of the atomic planes, will give rise to line defects, known as dislocations. Fig. 2.75(a) shows two distinct stages in the slipping from A to D and the two dislocations are known as edge and screw dislocations.

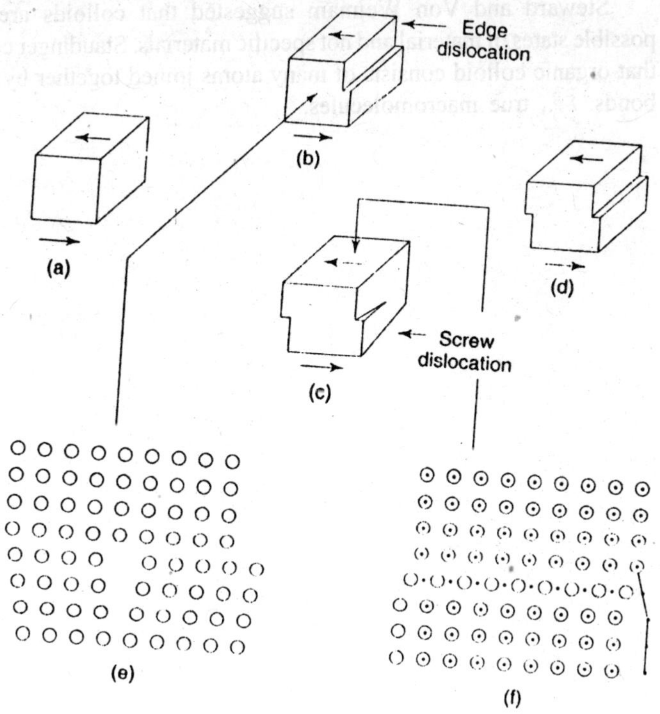

Fig. 2.75

The dislocation that contains an extra half plane of atoms at the slip plane is known as edge dislocation (Fig. 2.75). The line defect, from above and below the slip, generates a screw pattern (Fig. 2.75 f). In the figure the circles are above the slip plane and the dots representing atoms are below the slip planes. The distortions may be pure edge, pure screw or any possible combination of edge and screw components.

Hetero Chains

Styrene on heating changes from a clear liquid to a solid. Berthelot called this phenomenon as polymerization. He also observed that poly (styrene) on being heated at higher temperatures converts to styrene. This leads to miscelle theory of polymers. Staudinger classified colloids into colloidal dispersions miceller colloids (associated colloids) and colloidal molecules (macromolecular).

Steward and Von Weimam suggested that colloids are general possible states of material and not specific materials. Staudinger concluded that organic colloid consists of many atoms joined together by covalent bonds, *i.e.*, true macromolecules.

3

GASEOUS STATE

Although, the air was known to the primitive man, but the foundations of the modem concepts regarding the gaseous state were laid in the seventeenth century. In 1654, Otto von Guericke performed his famous *Magdebur hemisphere* experiment to show that the air exerted pressure. The actual development of the modem the theories began with the work of E. Torrcelli (Italy) in 1643. Torricelli measured the atmospheric pressure with the help of a barometer. In this chapter we discuss the physical behaviour of gases in terms of certain basic laws.

PHYSICAL CHARACTERISTICS OF GASES

All gases show some common characteristics. These are described below :

(i) Gases maintain neither the volume nor shape, and completely fill the container in which they are introduced.

(ii) Gases expand appreciably on heating.

(iii) Gases are highly compressible, that is, when the pressure is increased, the volume of a gas decreases.

(iv) Gases diffuse rapidly.

(v) All gases, except a few, are colourless. The following gases show characteristic colours, such as,

1. Nitrogen dioxide	1. Reddish-brown
2. Iodine	2. Violet
3. Bromine	3. Reddish-brown
4. Chlorine	4. Greenish-yellow
5. Fluorine	5. Greenish-yellow

(vi) Gases exert pressure equally in all the directions.

MEASURABLE PROPERTIES OF GASES

The behaviour of gases can be described in terms of certain parameters. These are described below.

Mass and Amount

The mass of a gas can be easily determined by weighting the container containing the gas and then emptying the container by taking out the gas and weighting the empty container again the difference between the two weights give the mass of the gas.

In chemistry, according to the IUPAC recommendations, the amount of a substance is expressed in terms of the number of moles (n). The number of moles can be obtained from the mass of the gas-by using the relationship,

$$\text{No. of moles (n)} = \frac{\text{Mass of the gas in gram (m)}}{\text{Molar mass of the gas in gram per mole (M)}}$$

Volume (V)

Volume of the as is equal to the volume of its equation. Volumes can be expressed in litre (L), millilitre (mL), or cubic centimetre (cm^3). The SI unit of volume is cubic metre (m^3). In common practice, however smaller units such as cubic decimetre (dm^3) and cubic centimetre (cm^3) are more frequently employed. These are related to each other as,

$$1 \text{ m}^3 = 10^3 \text{ dm}^3 = 10^6 \text{ cm}^3$$

The commonly used unit, litre (L) has now been redefined so that,

$$1 \text{ L} = 1 \text{ dm}^3$$

and $$1 \text{ mL} = 1 \text{ dm}^3$$

The volume (V) of any gas depends upon its amount, temperature and pressure. Mathematically, volume of a gas is a function of the amount, temperature, and pressure, *i.e.*,

$$V = f \text{ (amount, temperature, pressure)}$$

or $$V = f (n, T, P)$$

The instrument used for the measurement of the pressure of a gas is called a manometer. It simply consists of a U-shaped tube containing mercury usually. One limb of the tube is longer than the other. Two types of manometers are used. These are:

(i) Those in which the longer limb is closed (Fig. 3.1 .b)

(ii) Those in which the longer limb is open (Fig. 3.1 c)

Closed limb manometer is used only for gases at pressures less than the atmospheric pressure. The open limb manometer is used for all cases. In a case when the gas pressure is greater than atmospheric pressure, mercury stands at a higher level in the longer limb. In such a case, the difference of levels is added to the atmospheric pressure to get the pressure of the gas.

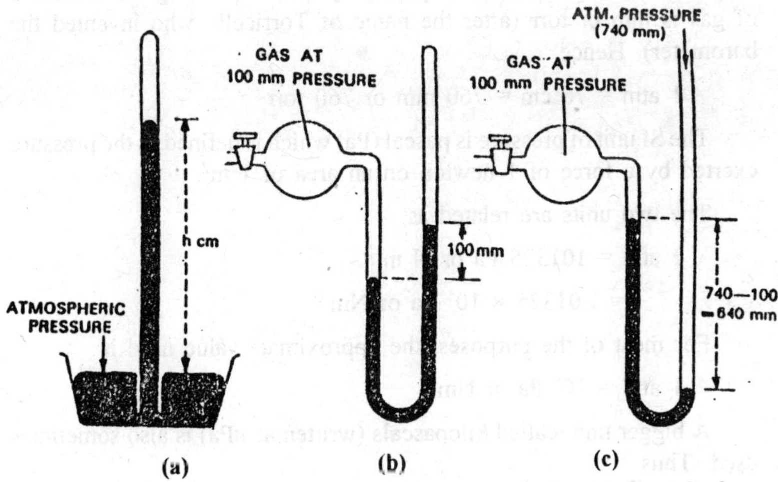

Fig. 3.1 : (a) Barometer (b) Closed limb manometer (c) Open limb manometer.

As pressure is force per unit area, the pressure obtained in terms of the height of the mercury column can be converted into force per unit area as follows:

Suppose height of the mercury column	= h cm
Area of cross-section of the tube	= Acm^2
∴ Volume of the mercury column	= A × hcm^3
If density of mercury at room temp	= ρ g cm^{-3}
then mass of the mercury column	= A × h x ρ gram

$\therefore$ weight of the mercury column $= (A \times h \times \rho) \times g$

where g if, the Gaseous State due to gravity.

This weight of the mercury column is the force acting on A cm^2. Hence.

$$\text{Pressure (P)} = \frac{\text{Force}}{\text{Area}} = \frac{A \times h \times \rho \times g}{A} = h\rho g$$

A *standard or normal atmospheric pressure* is defined as the pressure exerted by a mercury column of exactly 76 cm at 0°C. This is the pressure exerted by the atmosphere at the sea level.

The smaller unit commonly employed in expressing the pressures of gas is mm or torr (after the name of Torricelli, who invented the barometer). Hence

1 atm = 76 cm = 760 mm or 760 torr

The SI unit of pressure is pascal (Pa) which is defined as the pressure exerted by a force of 1 newton on an area of 1 m^2.

The two units are related as

1 atm = 101325 Pa or N m^{-2}

= 1.01325×10^5 Pa or Nm^{-2}

For most of the purposes, the approximate value used is

1 atm = 10^5 Pa or Nm^{-2}

A bigger unit, called kilopascals (written as kPa) is also sometimes used. Thus

1 atm = 10^2 kPa

Measurement of Temperature : Temperature is a measure of the extent of hotness or coldness of a body. The measurement of temperature is based upon the principle that substances expand on heating. The most common substance whose expansion is made use of in the measurement of temperatures is 'mercury'. There are three different scales on which the temperatures are measured. These are

(i) Centigrade or Celsius scale (after the name of Anders Celsius)

(ii) Fahrenheit scale (after the name of Daniel Fahrenheit, a German instrument maker)

(iii) Kelvin scale (after the name of Lord Kelvin)

The Celsius scale is based upon taking the freezing point of water as 0°C and the boiling point of water as 100°C at normal atmospheric pressure (*i.e.* at sea level) and then dividing the range into 100 equal parts so that each part represents 1°C.

The Fahrenheit scale is based upon taking the freezing point of water as 32°F and the boiling point of water as 212°F and dividing the range into 180 equal parts. Thus, A large number of chemicals that go to form the basis of bigger molecules required for the synthesis of high polymers exist in gaseous state at the room temperature and atmospheric pressure, *e.g.*, acetylene, ethylene, propylene, 1-Butene, etc. In order to follow the mechanism of the formation of these bigger molecules and their subsequent polymerization, it is essential that we may not only study the characteristics and behaviour of these gases, but also the physical laws that govern their behaviour.

One of the main objectives of modern chemistry is to relate the properties of a large collection of molecules which constitute the matter in bulk (macroscopic properties) to the properties of individual atoms and molecules (microscopic properties). Inspite of the fact that the particles of a gas are always in random motion, a relation between microscopic and macroscopic behaviour is simpler in gases than in solids or liquids. The molecules of a gas, even at ordinary temperature and pressure, move with considerable kinetic energy, but they are least affected by the forces of attraction and repulsion. The gases do not possess volume of their own but have attributes like touch and sound. They possess properties like mass, compressibility, refractive index, density and even electrical conductivity at very low pressure when the molecules get charged. Their behaviour has been found to depend on the population density of the molecules in a specified space.

TYPES OF GASES

There are two types of gases viz. *uniform* and *non-uniform*.

A gas need not normally have its mass, pressure, volume and temperature the same at all the places. Generally, it is a mixture of several elements as air is a mixture of oxygen, nitrogen, water vapour, carbon dioxide, etc. Its composition vary from place to place. Even in a chemically pure substance, *e.g.*, nitrogen, it may be considered that either the mass or volume or temperature or pressure or all of them may vary from place to place. Such gases are called non-uniform gases.

It is always possible to find out a small volume element where all the above measurable properties are constant. Such gas element are called uniform gases.

PROPERTIES OF GASES

There are two classes of properties viz. extensive and intensive.

The properties that depend on the amount of the gas like volume, mass, etc., are called extensive properties while in a uniform gas the properties that are independent of the amount of the gas, *e.g.*, pressure, temperature are called intensive properties.

It will not be possible to define a uniform gas. It is one that has the value of each of its intensive property the same irrespective of the region of the gas.

STATE OF A GAS

Each condition of a gas, *i.e.*, mass, volume, temperature, pressure, etc., which is completely distinguishable from any other condition is called as the state of the gas. The above measurable properties are called the state variable of the gaseous system. The state of a system is a function of its state variables.

Mass

The gas is composed of innumerable molecules each having a mass of the order of 10^{-24}g. So all these molecules collectively contribute towards the mass of the gas.

Occupies Space

The space occupied by all molecules of the gas is very small as compared to the space occupied by the gas. It is roughly of the order of 1000th part of the space occupied by the gas. The remaining space at any instant is vacant, and as such it occupies a volume. In this context it would not be wrong to approximate the molecule as a point in the whole space. The volume is measured as millilitres.

Motion

At any temperature above absolute zero, the molecules of a gas are in motion. At room temperature the molecules of air move at a speed of 5×10^4 metres per second (1100 miles per hour), so they have fluidity.

Diffusivity

The process by which a gas moves from a high pressure through a porous wall is known as diffusion. The molecules of a gas are moving at a fast speed and do not occupy any appreciable volume by themselves, hence, they will not oppose expansion. So they will have very high diffusivity.

Isotropic

The molecules in a gas, occupying insignificant space, are moving randomly with high speed and as such do not have any effective attractive or repuslive force between each other with the difference that if one molecule moves in a stipulated direction then it must correlate with the other, moving simultaneously in the opposite direction. Thus, they will occupy the space of the container and as such will not have a volume of their own and can be easily added or separated if no change of composition is to be achieved. They are mutually soluble as there is too much vacant space in between them. So its movement appears to be equal in all the directions. Since they have the same movement in all directions, they are known as isotropic.

Stationary State

Any change in the motion of one molecule is correlated with the compensatory change in the other molecule. Therefore, unless external factor is introduced the state of the gas remains stationary.

Stationary Energy

The molecules of the gas collide with the walls of the container and rebound. These collisions are assumed to be perfectly elastic and as such exert a pressure on the walls of the container but do not lose energy on rebounding from it. Therefore, at a given temperature the total energy of the system remains constant.

LAWS OF NEWTON

The olecules of the gas change direction on striking against the wall of the container. In all these cases they are guided by the Newton's laws of motion. These laws provide equation of state of uniform gases at low pressures.

The behaviour of a uniform pure gas like hydrogen, oxygen, nitrogen, carbon dioxide, isopentane, etc., is governed by pressure and temperature as intensive variable and mass and volume as extensive variable. This can be mathematically expressed by the equation:

$$F(m, V, P, t) = 0 \qquad ...(1)$$

Since V and m have extensive nature

$$V = V(m, P, t) \qquad ..(2)$$

$$= mV(1, P, t) \qquad ...(3)$$

Eqn. 3 represents a function depending on P and t as variables and refers to a unit mass which can be written as

$$V = m\overline{\upsilon}\ (P, t) \qquad ...(4)$$

where $\overline{\upsilon}$ is the volume for a unit mass and is known as specific volume of the gas.

Table 3.1 : Molar and specific volumes for certain gases at 1 atm 0°C

Gas	H_2	He	N_2	O_2	CO	AH_3	CH CH	CH_2 CH_2
Molar volume 1 litre	22.432	22.396	22.403	22.392	22.400	22.263	22.085	22.246
Specific volume I litre	11.13			0.723	0.800			

BOYLE'S LAW

Boyle in 1661 studied the effect of pressure on the volume of a fixed mass of air. Boyle's concept of air was as an elastic fluid and he called pressure as the spring of air.

Boyle observed that on the application of pressure to air at constant temperature, the volume of air decreased inversely with the pressure applied to it.

It was soon observed that other gases also showed inverse relationship between volume and pressure at constant temperature.

Fig. 3.2(a) shows smooth rectangular hyperbola expressing the inverse proportionality of P and V, *i.e.*, V tends to 0 as P tends to infinity or conversely. A better proof of Boyle's laws obtained in Figs. 3.2 (b) and 3.3, where straight line is obtained by plotting pvs 1/υ or PV vs P.

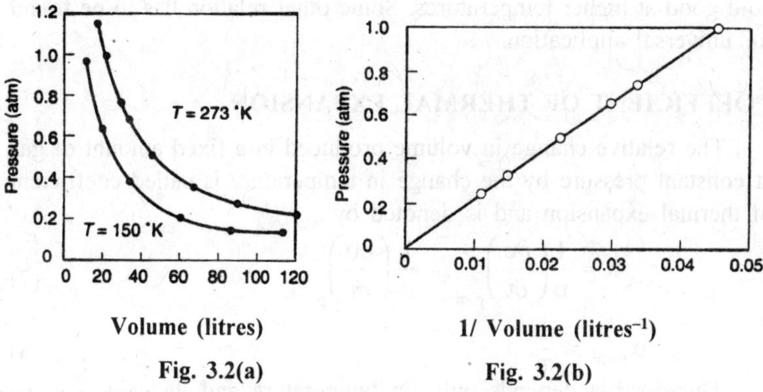

Fig. 3.2(a) Fig. 3.2(b)

The inverse relationship can be written as:

$$PV = \text{Constant}$$

It can be better expressed as

$$PV = K_1 (n, T)$$

where K_1 is constant whose value depends upon temperature (T) and the number of moles of gas (n).

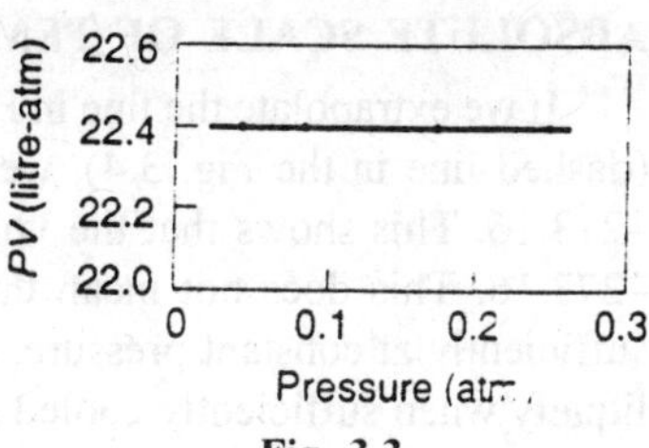

Fig. 3.3

Example:

One mole of 2-fluoropropane occupies 25.54 litres at 703.6mm pressure and 25°C.

Solution:

Calculate its pressure when it is compressed to 7.055 litres at 25°C.

$$P_1V_1 = P_2V_2$$

$$P_1 \times 7.0551 = 703.6 \text{ mm} \times 25.541 \quad ...(5)$$

$$P_1 = \frac{703.6\text{mm} \times 25.541}{7.0501} = 2546.7 \text{ mm} \quad ...(6)$$

Table 3.2 : PV value for acetylene at 0°C

P atm	0.5	1.0	2.0	4.0	8.0
PV atm	1.0057	1.0000	0.9891	0.9780	0.9360

It is evident from the above that the Boyle's law holds good only at low pressures. Similarly, it was observed that Eqn. 6 was found to hold good at higher temperatures. Some other relation has to be found for universal application.

COEFFICIENT OF THERMAL EXPANSION

The relative change in volume produced in a fixed amount of gas at constant pressure by the change in temperature is called coefficient of thermal expansion and is denoted by

$$\alpha = \frac{1}{\upsilon}\left(\frac{\partial \upsilon}{\partial t}\right)_{p,m} = \frac{1}{\upsilon}\left(\frac{\partial \upsilon}{\partial t}\right)_{p} \qquad ...(7)$$

$$\alpha_{ideal} = \frac{1}{T} \qquad ...(8)$$

Therefore, a depends only on temperature and its value can be obtained by plotting isobars between V vs T (Fig. 3.4).

ABSOLUTE SCALE OF TEMPERATURE

If we extrapolate the line in Fig. 3.4 beyond the actual measurements (dashed line in the Fig. 3.4), we find that it cuts the horizontal axis at –273.16. This shows that the volume of the gas would become zero at –273.16. This does not mean that the gas will disappear if we cool it sufficiently at constant pressure. What actually happens is that all gases liquefy when sufficiently cooled at constant pressure and they no longer obey Boyle's law. It only means that an ideal gas is a hypothetical gas which does not liquefy and shrinks until it finally occupies no volume.

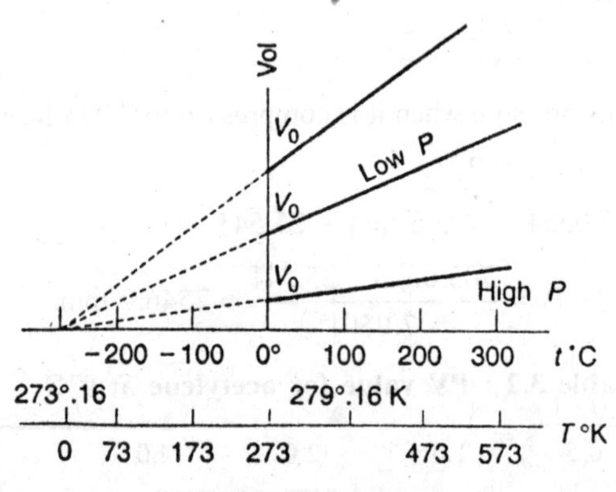

Fig. 3.4

This temperature of –273.16 is called absolute zero of temperature because any temperature lower than this would correspond to a negative volume of the gas. This temperature has been approached experimentally, but has never been reached. At least we have approached within about 0.0001 of absolute zero.

The value of α for an ideal gas is 0.36 per cent per degree. Fig. 3.4. Since t is a linear function of volume Va.t constant pressure P, the value of r is given by:

$$t = \frac{V - V_0}{V_{100} - V_0} \times 100 \quad ...(9)$$

where V, V_0 and V_{100} are volumes t°C, 0°C and 100°C respectively.

• The ideal gas can be used to make an ideal gas thermometer. Since t and V are always positive. Lord Kelvin proposed a new scale of temperature in which the zero point corresponds to the value of temperature at which the volume of an ideal gas becomes zero. It is now known as Kelvin scale of temperature where T° = t° + 273.16°. Generally, a value of 273° is used instead of 273.16° and F denotes the absolute temperature.

CHARLES AND GAY LUSSACS LAW

Charles and Gay Lussac in 1802 studied the effect of temperature on volume at constant pressure and concluded that all gases have the same coefficient of expansion at the same temperature provided the pressure and amount of gas are kept constant.

$$V = V_0 + \left(\frac{\partial \upsilon}{\partial t}\right)^t \text{ (n, P constant)} \quad ...(10)$$

In Fig. 3.4 the intercept at the vertical axis is Vg which is the volume at t = 0° and slope of the line is $\frac{\partial \upsilon}{\partial t}$ which is α.

$$V_t = V_0 + V_0 \alpha t$$

$$= V_0\left(1 + \frac{t}{273.16}\right) = V_0\left(\frac{t + 273.16}{273.16}\right) = V_0 \frac{T}{T_0}$$

$$\frac{V_1}{T} = \frac{V_0}{T_0} \quad ...(11)$$

Since both VQ and a are positive.

$$\frac{V}{V_0} = \frac{T}{T_0} = K_1(t) \quad ...(12)$$

where K_1 is a constant. By plotting K(t) vs T (Fig. 3.5), a straight line was obtained showing a linear relation. It is mathematically represented as:

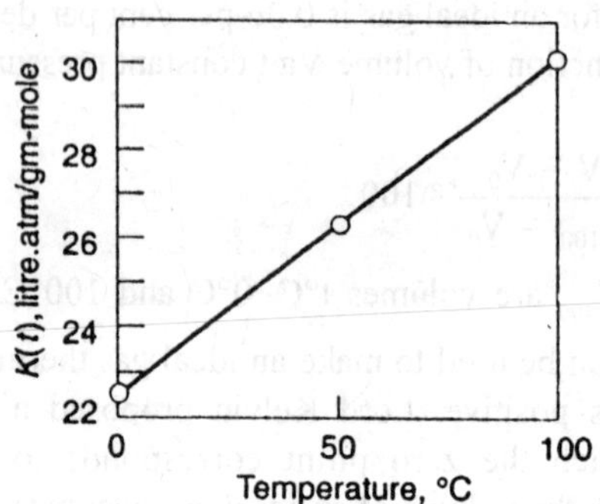

Fig. 3.5

$$Kt\ (t) = RT \qquad ...(13)$$

where R is a new universal constant.

The equation of state can now be written as:

$$P\overline{V} = K_1\ (T) = RT \qquad ...(14)$$

where $\overline{V}$ is wv, *i.e.*, volume per gram mole of the gas known as Molal volume.

AVOGADRO'S HYPOTHESIS

In 1811, Amedeo Avogadro suggested that equal volumes of all gases contain equal number of molecules under the same conditions. This means that volume is proportional to the number of molecules.

Cannizzaro in 1860 on the basis of the above proved that water is H_2O and not HO as was believed earlier. Avogadro defined a new term mole of a substance. The number of grams equal to its molecular weight is known as Mole. One mole of the gas at the same temperature and pressure will contain the same number of molecules given by (6.023 × 1023). This figure is called Avogadro's number and applies to one mole of any compound. Avogadro's hypothesis in equation form can be written as:

$$V = K_2(t, P)n \qquad ...(15)$$

where K_2 is a constant.

EQUATION OF STATE

Bernauli independently arrived at a similar equation of state. Let V_1, P_1 and T_1 be the volume, pressure and temperature of a mass of gas respectively. Let us do some work.

(1) At constant temperature, change the pressure P_1 to P_2, so that the volume V\1 becomes V_x

$$V_x = V_1 \frac{P_1}{P_2} \text{ (Boyle's law)} \qquad ...(16)$$

(2) At constant pressure P_2, change the temperature T_1 to T_2, so that volume changes from V_x to V_2

$$V_x = V_2 \frac{V_1}{V_2} \text{ (Charle's law)} \qquad ...(17)$$

$$V_1 \frac{P_1}{P_2} = V_2 \frac{T_1}{T_2}$$

or
$$\frac{P_1 V_1}{T_1} = \frac{P_2 V_2}{T_2} = \frac{PV}{T} = \text{constant} \qquad ...(18)$$

Again
$$d\upsilon = \left(\frac{\partial \upsilon}{\partial P}\right)_T P + \left(\frac{\partial \upsilon}{\partial t}\right)_P T \qquad ...(19)$$

Now
$$\left(\frac{\partial \upsilon}{\partial P}\right)_T = -\frac{K_1}{P_2} \text{ (Boyles law)} \qquad ...(20)$$

$$\left(\frac{\partial \upsilon}{\partial T}\right)_P = K_2 \quad \text{(Charles law)} \qquad ...(21)$$

$$dv = \frac{K_1}{P_2} dp + K_2\, dT \qquad ...(22)$$

$$= \frac{V}{P} dp + \frac{V}{T} dT$$

or
$$\frac{v}{V} = \frac{P}{P} + \frac{T}{T}$$

on integration

$$\frac{PV}{T} = \text{constant} \qquad ...(23)$$

The equation of state for an ideal gas can be written as

$$PV = nRT \qquad ...(24)$$

Expressing the equation in terms of density and mass,

$$P = \frac{nRT}{V} = \frac{m}{M}\frac{RT}{V} = \frac{\rho RT}{M} \quad ...(25)$$

where n, m, M and ρ are the number of moles, mass of the gas, molecular mass and density of the gas respectively.

The gas that follows the above equation is called an Ideal gas.

Gas Constant

The ideal universal gas constant R has a value of $\frac{P\overline{V}}{T}$ for one mole of a gas and is independent of the nature of the gas. One mole of a gas occupies a volume of 22.414 litres at 1 atmosphere pressure and 273°K.

$$= \frac{P\overline{V}}{T} = \frac{1 \text{ atom} \times 22.414 \text{ litres}}{1 \text{ mole} \times 273^\circ \text{K}}$$

$$= 0.0821 \text{ litre atm mole}^{-1} \text{ }^\circ\text{K}^{-1} \quad ...(26)$$

R has the dimensions of energy per degree per mole.

Table 3.3 : Values of ideal gas constant R in different units

$R = 0.0821 \text{ 1 atm K}^{o-1} \text{ mole}^{-1}$
$= 82.057 \text{ ml atm K}^{on} \text{ mole}^{-1}$
$= 62.3 \text{ 1 – mm Hg K}^{o-1} \text{ mole}^{-1}$
$= 8.3143 \text{ J}^\circ\text{K}^{-1} \text{ mole}^{-1}$
$= 1.987 \text{ carl }^\circ\text{K}^{-1} \text{ mole}^{o-1}$

Boltzmann denoted the gas constant per molecule as K.

$$K = \frac{R}{W} = \frac{8.314 \times 10^7 \text{ erg deg}^{-1} \text{ mole}^{-1}}{6.023 \times 10^{23}}$$

$$= 1.3806 \times 10^{-16} \text{ erg deg}^{-1} \text{ mol}^{-1} \quad ...(27)$$

The average kinetic energy

$$\overline{E} = \frac{3}{2} KT$$

$$= 6.17 \times 10^{-14} \text{ erg at } 25^\circ\text{C} \quad ...(28).$$

MIXTURE OF GASES, DALTON'S LAW OF PARTIAL PRESSURES

So far we have considered the relation between P, V, T, and n for a gaseous system in which only one component is present. If several substances are present in a gaseous mixture, we can still use the gas laws by taking into account the presence of different substances. The relationship between P, V, T and n in the mixture of gases can be illustrated by the experiment.

Suppose the sample of nitrogen is pumped into the first vessel and its pressure is found to be P_1 cm Hg; a sample of oxygen is pumped into the second vessel and its pressure is found to be P_2 cm Hg. If both the samples are pumped into the third vessel, the pressure is observed to be the sum of P_1 and P_2. This shows that each component independently contributes to the total pressure provided they do not react.

This behaviour of a mixture of two gases was first expressed in a generalized form in 1807 by John Dalton and is known as Dalton's law of partial pressures. This law states that the total pressure exerted by a mixture of non-reacting gases is equal to the sum of the pressures which each component would exert if placed separately into the container:

$$P_t = P_1 + P_2 + P_3 + \ldots \text{ (T, V = constant)} \qquad \ldots(29)$$

where P_t is the total pressure and P_1; P_2, P_3, are partial pressures of components 1, 2, 3 ..., etc. In a concise form, we can write the above expression as,

$$P_t = \sum_i P_i \text{ (V and T constant)} \qquad \ldots(30)$$

where the symbol $\sum_i$ stands for the summation over all the components present in the mixture.

The partial pressure P_i of component; is defined as the pressure that the gas would exert if it was present alone in the same volume and at the same temperature. Thus, if a container of volume contains n_1 moles of gas 1, n_2 moles of gas 2, etc., then the partial pressures of gases 1, 2, etc. are

$$P_1 = \frac{n_1RT}{V}, P_2 = \frac{n_2RT}{V}$$

The total pressure of the system can be written as

$$P_t = \sum_i P_i \frac{RT}{V} \sum_i n_i = n_t \frac{n_2RT}{V} \qquad \ldots(31)$$

where n_t is the total number of moles of gas present in the mixture, $n_t = n_1 + n_2 + n_3 + \ldots = \sum_i n_i$. This equation is an even more general form of the ideal gas law, valid for mixtures as well as for pure gases;

it combines in one equation, the laws of Boyle, Charles, Avogadro and Dalton.

For mixtures of gases, it is sometimes convenient to express the law of partial pressures in terms of mole fractions of the various components of the mixture. The mole fractions, of a component i is defined by $x_i = n_i/n_t$, where n_i is the number of moles of component i and n_t is the total number of moles. The partial pressure P_i of component is then given by

$$P_i = x_i P_t \qquad ...(32)$$

Remember that $\sum_i x_i = 1$, so that $\sum_i P_i = P_t \sum_i x_i = P_t$

Equation 32 represents Raoults law.

A law similar to Dalton's law is Amagat's law of partial volumes. According to this law, $V_t = \upsilon_1 + \upsilon_2 + \upsilon_3 ...$ at constant temperature and pressure. Here V_t represents the total volume and $\upsilon_1, \upsilon_2 ...$ the partial volumes. Thus, when two or more gases are mixed at constant volume and temperature their pressures are additive; when they are mixed at constant pressures and temperature, their volumes are additive.

$$V_1 = n_i \frac{RT}{P} \qquad ..(33)$$

or

$$W = \frac{m_1}{P_1} \frac{RT}{V} \qquad ...(34)$$

The application of Dalton's law is illustrated by the following examples.

Example:

12.59 × 10⁻³ moles per litre of N^ when heated to 25°C exerts a pressure of 0.3942 atmos. The gas was found to contain a mixture of N_2O_2 and NO_2. Calculate the average molecular weight and partial pressures due to each.

Solution:

$$M_{Av} = \frac{(12.59 \times 10^{-3})\ 92.2 \times 8.2 \times 10^{-2} \times 298}{0.3942}$$

$= 71.78$ g mole^{-1}.

$^{46x}NO_2 + {}^{92}N_2O_4 = 71.78$ g mole^{-1}

$46 - {}^{46x}N_2O_4 + {}^{92x}N_2O_4 = 71.78$ g mole^{-1}

$$1 + {}^{x}N_2O_4 = .56 \text{ or } {}^{x}N_2O_4 = 0.56$$

$$P_{N_2O_4} = (0.56)\,(0.3942) \text{ atm} = 0.0221 \text{ atm}$$

$$P_{NO_4} = (0.3942 - 0.0221) \text{ atm} = 0.3721 \text{ atm}$$

Specific Volume

The specific volume serves as a very important quantity in calculating the molecular weight of a gas.

$$PV = nRT = \frac{m}{M} RT$$

$$M = \frac{m}{V}\,\frac{RT}{P} = \rho\,\frac{RT}{P} \qquad \text{...(35)}$$

$$V_{ideal} = \frac{8.2 \times 10^{-2} \text{ 1 atm/mole} \times 273° \text{K}}{1 \text{ atm}}$$

$$= 22.4151/\text{g mole} \qquad \text{...(36)}$$

Fig. 3.6 gives a plot between d/p and P at 0°C for argon

As P approaches zero d/p is 1.7826 g/1 atm

The molecular weight of Argon

$= 8.2057 \times 10^{-2}$ 1 atm/deg mole × 273 °K × 1.7826 g/1 atm

= 39.955 g/mole.

GRAHAM'S LAW

Gases have a remarkable tendency to spread throughout the entire space available to it. When a small quantity of hydrogen sulphide gas is released in a laboratory its smell is felt soon in all parts of the laboratory. The gas has diffused through the atmosphere and distributed itself throughout the room. This property of a gas which involves the movement of the gas molecules through gases of another kind is called diffusion.

Diffusion phenomenon may easily be illustrated with the help of an experiment. Introduce simultaneously two cotton plugs, one soaked in concentrated hydrochloric acid and the other in liquor ammonia at the ends of a glass tube (1 cm diameter). Close the ends by cork to prevent air draughts. In a short while we find the formation of a white ring in the tube at a place closer to the end where HC1 cotton wad is introduced. White cloud is ammonium chloride which is formed when HC1 and NH^3 move towards each other. The fact that the cloud is nearer to HC1 implies that HC1 molecules moved relatively slowly. We know that a molecule of HC1 is heavier compared to that of ammonia. From this it may be concluded that lighter gases move faster than heavier gases under identical conditions of temperature and pressure.

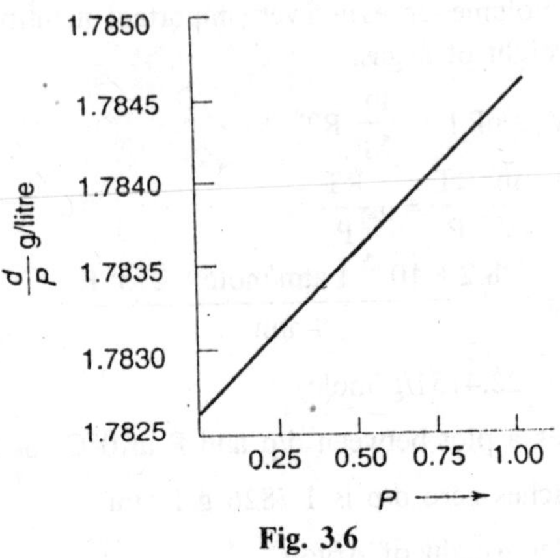

Fig. 3.6

Diffusion phenomena can be studied under controlled experimental conditions and is described quantitatively by Graham's law which states, under the same conditions of temperature and pressure the rate of diffusion of a gas is inversely proportional to the square root of its molecular mass.

Mathematically the law can be expressed as,

$$r \propto \frac{1}{(M)^{1/2}}$$

or

$$r = \frac{\text{const}}{(M)^{1/2}}$$

$$\frac{r_1}{r_2} = \left(\frac{M_2}{M_1}\right)^{1/2} = \left(\frac{\rho_1}{\rho_2}\right)^{1/2} \qquad ...(37)$$

where r_1 and r_2 are the rates of diffusion of gases 1 and 2 and M_1 and M_2 are their molecular masses and ρ_1 and ρ_2 are respective density of the gas. Since it is rather difficult to deal with absolute rates of diffusion of gases, attention is usually confined to the relative rate of diffusion of gases. The law is useful for the determination of moloculer masses of gases, and is utilized for the separation of U^{238} and U^{235}. Their fluorides have been made to diffuse through a porous solid where $U^{235}F_6$ diffuses more rapidly than heavier $U^{238}F_6$. The enrichment factor was found to be 1.004 at each cycle. It should, however, be remembered that this law is true only for gases diffusing under low pressure.

KINETIC MOLECULAR GAS MODEL

In the preceding sections of this chapter, we examined the physical behaviour of gases from an empirical point of view and paid no attention to questions like why gases occupy the entire space available to them, why they are highly compressible, how gases exert pressure, why different gases diffuse at different rates, etc. Since it is not possible to deduce the nature of gases from the measured properties alone, it becomes necessary to look for a model to account for the observed facts. In order that the model is simple, certain assumptions will have to be made about the nature of gases. We shall now examine such a model first by discussing the validity of assumptions and later by comparing its predictions with observed facts.

Assumptions of Kinetic Theory

The following assumptions are made to describe an ideal gas :

(i) A gas consists of a large number of particles (atoms or molecules) that are so small and so far apart (on the average) that the volume of the molecules is negligible compared to the empty space between them. This assumption seems to be justified in view of the great compressibility of the gas. Calculations also show that about 99.9% of the total volume in nitrogen gas at STP is empty space. The average spacing between molecules of nitrogen at STP is worked out to be about ten times the molecular diameter. When the gas is compressed, the average spacing between molecules is reduced.

(ii) There are no attractive forces between molecules and the molecules are completely independent of one other. The support for this assumption comes from the observation that even highly compressed gases expand spontaneously to occupy all the space available. This suggests that there is no appreciable interaction between molecules.

(iii) The molecules of a gas move randomly in straight lines colliding with one another and also with the walls of the container. The pressure exerted by a gas is the result of collisions of the molecules with the walls of the container. That particles of a gas are not at rest but in a state of continuous motion (called thermal agitation) cannot be directly observed because molecules are too small to be seen individually even through the best microscope. However, movement of particles of size in the neighbourhood of 1 p. diameter (*e.g.*, particles of smoke floating in air) can be observed microscopically. This phenomenon is called Brownian motion (after the British botanist who first observed the irregular motion of pollen particles suspended in water). It is found that the smaller the particles, the more violent is the irregular motion; also, the higher the temperature the more vigorous is the movement of the particles. In the absence of attractive forces between molecules, it is reasonable to assume straight line random motion for the molecules which are much smaller in size. Molecules, however, change the path as a result of collision with other molecules or the walls of the container.

(iv) In these collisions, there is no net loss of kinetic energy, although, there may be a transfer of energy between the partners in the collisions (elastic collisions). If there were loss of kinetic energy in collisions (inelastic collisions), the motion of molecules would eventually stop and the pressure and temperature would drop to zero. Since the pressure of a confined gas does not change with time (unless we heat or cool it), we conclude that a gas exerts pressure through elastic collisions with the wall.

(v) At any particular instant, different molecules in a sample of gas have different speeds and hence, different kinetic energies, but the average kinetic energy of the molecules is directly proportional to the absolute temperature.

Root Mean Square Velocity

The molecules of a gas in an enclosed space suffer continuous changes of momentum through collision with other molecules. As a result of such collision any individual molecule will acquire velocities ranging from zero to infinity. It is quite probable that any molecule in a given time may exhibit an average velocity zero as there are equal probability of its movement in any direction (both positive and negative) but the average of its square velocities shall always be positive.

$$\sqrt{\overline{u^2}} = \sqrt{\frac{u_1^2 + u_2^2 + u_3^2 + u_4^2 \ldots u_n^2}{n}} \quad \ldots(38)$$

Distribution of Speed

The assumption that at any instant different molecules have different speeds and hence, different energies is valid in view of the great number of molecular collisions. As molecules collide with each other, we would expect the speed of the molecules to change. All molecules in a container will, therefore, not have the same speed at any one time. Even if the molecules had the same initial speed, the effects of collisions would soon disrupt such uniformity. The speed of each gaseous molecule probably changes billions of times during a second and we have no way of knowing the speed or kinetic energy of each individual molecule. Instead, we shall make a statistical approach to predict as to how many of the molecules have a particular speed and kinetic energy. We shall now examine the distribution of molecular speeds briefly.

In order to understand the distribution of molecular speeds, it may be helpful to look at a familiar distribution. As an example, we consider the distribution of performance of students in a class test. Let us suppose that there are 100 students and we have information about the marks (percentage-wise) obtained by each student. It is difficult to obtain a reasonably accurate idea of how the marks are distributed among the students by looking at the marks of each student. For the purpose of obtaining a better idea of how the marks are distributed in the class we study the performance distribution of the students by grouping together all students with marks within a certain range. Let this be 10. Then all students having marks ranging between, say 10-20%, 20-30%, etc., will be grouped together as shown in Table 3.4, and plotted in Fig. 3.7. It is obvious that not all the students have the same marks and the students in the class are 'distributed' over this range of marks. The height of each rectangle refers to the number of students having marks within the range indicated by the base of the rectangle.

Table 3.4

Marks (%)	*Number of students*
10-20	1
20-30	4
30-40	10
40-50	20
50-60	40
60-70	20
70-80	5

A more precise distribution can be obtained if a smaller value of 'marks range' is adopted and the corresponding distribution determined. If the range is very small, and the number is very large, we have an outline like the smooth curve given in the Figure 3.7.

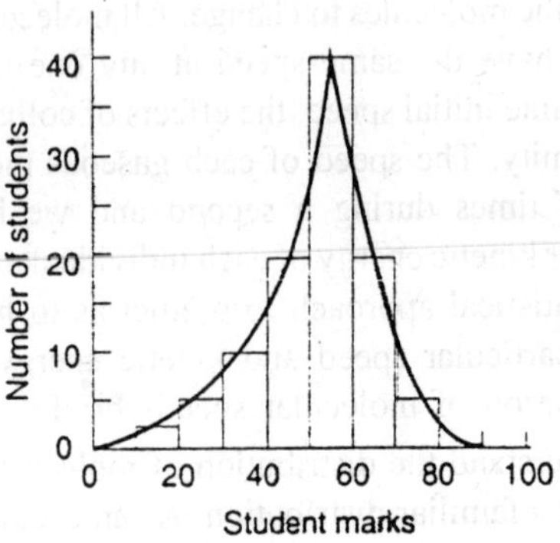

Fig. 3.7

In the case of gases. Maxwell and Boltzmann independently predicted theoretically, the shape of the distribution curve of molecular speeds in a sample of gas at a given temperature, assuming that the molecular collisions are completely random and involve all possible values for the exchange of kinetic energy. It is only the total kinetic energy of a gas that stays constant at a fixed temperature, but not individual kinetic energies of molecules. Fig. 3.8a shows an experimental arrangement for obtaining data on molecular speeds. The data obtained from such experiments fit the theoretical curves within the limit of experimental errors (Fig. 3.8b). Two discs D_1 and D_2 rotate rapidly on the same axle. These discs are in an evacuated chamber and rotate in front of an oven containing molten tin. Some tin vapourizes and steams out of the small opening in the oven, striking the first rotating disc, D_1. Once during every rotation the slot in this disc is in line with the hole of the oven. Then a small sample of tin molecules can pass through the slot and move on towards the second disc D_2. The fast moving molecules lead the way. Slower moving molecules begin to lag behind. As the molecules reach the second disc they condense on it. Since this disc is also rotating, fast moving molecules would condense on Sections B, C and D. Slow-moving molecules take a longer time to reach the second disc. The speed of the rotating disc is adjusted so that the slowest molecules condense on Sections I, J and K. Many different groups of tin molecules pass

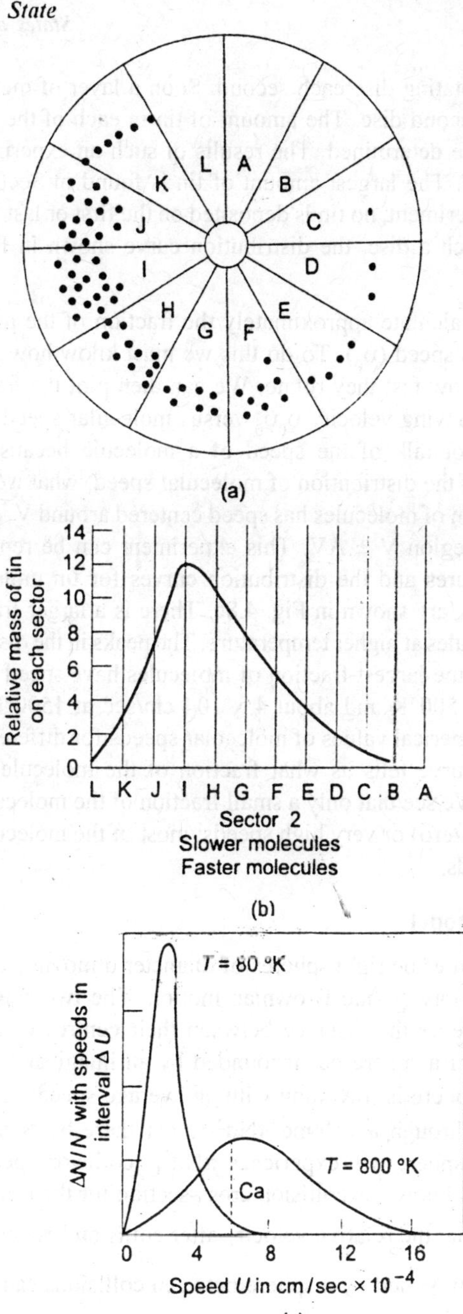
L
A
K
B
J
C
I
D
H
E
G
F
(a)
Relative mass of tin on each sector
14
12
10
8
6
4
2
0
L K J I H G F E D C B A
Sector 2
Slower molecules
Faster molecules
(b)
T = 80 °K
T = 800 °K
Ca
ΔN/N with speeds in interval ΔU
0
4
8
12
16
Speed U in cm/sec × 10^−4
(c)

Fig. 3.8

through the first rotating disc each second. Soon a layer of metallic tin builds up on the second disc. The amount of tin in each of the sections A through L can be determined. The results of such an experiment are shown in Fig. 3.8a. The largest amount of tin is found in Section I, In our illustrative experiment, no tin is deposited on the first or last Sections A and L. From such a disc, the distribution curve shown in Fig. 3.8b can be derived.

We can also calculate approximately the fraction of the molecules having a particular speed (υ_0). To do this we must know how far apart the discs are and how fast they rotate. We can then plot the fraction of the tin molecules having velocity $\upsilon_0\upsilon_0$ versus molecular speed. Strictly speaking we cannot talk of the speed of a molecule because of the statistical nature of the distribution of molecular speed; what we can say is this much fraction of molecules has speed centered around V, *i.e.*, their speeds lie in the region V ± AV. This experiment can be repeated a{ different temperatures and the distribution curves for tin molecules at 500°K and 1500°K are shown in Fig. 4.8c. There is a larger fraction of fast-moving molecules at higher temperature. The peaks in the distribution curves tell us that the largest fraction of molecules have speed of about 2×10^4 cm/sec at 500°K and about 4 x 104 cm/sec at 1500°K. Table 4.5 gives some numerical values of molecular speeds for different gases. The distribution curve tells us what fraction of the molecules have a particular speed. We see that only a small fraction of the molecules have very low (close to zero) or very high speeds; most of the molecules have intermediate speeds.

Random Walk Model

Let the molecules be rigid spheres of diameter d moving at random at an average velocity $\overline{c}$ like Brownian motion. The two spheres will collide only whenever the distance between their centres will become d. Let the centre of a sphere be surrounded by an imaginary sphere of radius d. As the sphere is travelling with an average speed $\overline{c}$, in a unit time it will pass through a volume $\pi Nd^2\overline{c}$. Let there be N molecules per c.c. Then the sphere will experience $\pi Nd^2\overline{c}$ collisions per second. The quantity ndl is known as collision cross-section for the rigid sphere. It can be proved that the relative velocity after collision becomes $\sqrt{2\overline{C}}$.

The distance travelled by a sphere between collisions can be given by $\dfrac{1}{\sqrt{2}\,N\pi d^2}$ with velocity U_x, then this velocity U_x receives increments of velocity from random collision with other molecules. Feller has shown that the probability that a molecule will attain a velocity U_x can be given as:

$$p(U_x) = (2\pi\alpha^2)^{-1/2} \exp\left(U_x^2/2\sigma^2\right) \quad ...(39)$$

Where a is known as the standard deviation from the mean. We also know that in one dimension

$$\frac{1}{2}\text{Nm}\,\overline{V}^2 = \frac{1}{2}RT \quad ...(40)$$

or
$$\overline{V}^2 = \frac{R}{N}\frac{T}{m} \quad ...(41)$$

Since velocity can be both positive or negative

$$\sigma^2 = V^2 \quad ...(42)$$

Therefore, the probability that a molecule has a velocity between U_x and $U_x + dU_x$

$$P(U_x)\,dU_x = \left(\frac{m}{2\pi KT}\right)^{1/2} \exp\left(-mU_x^2/2\,KT\right) dU_x \quad ...(43)$$

If there are large number of molecules N, then the fraction

$$\frac{dN(x)}{N} = \left(\frac{m}{2\pi KT}\right)^{1/2} \exp\left(-mCUx^2/2\,KT\right) dU_x \quad ...(44)$$

Jeans has applied similar arguments to two dimensions where the simultaneous velocity components are U_x and U_y, then the velocity C regardless of the direction is

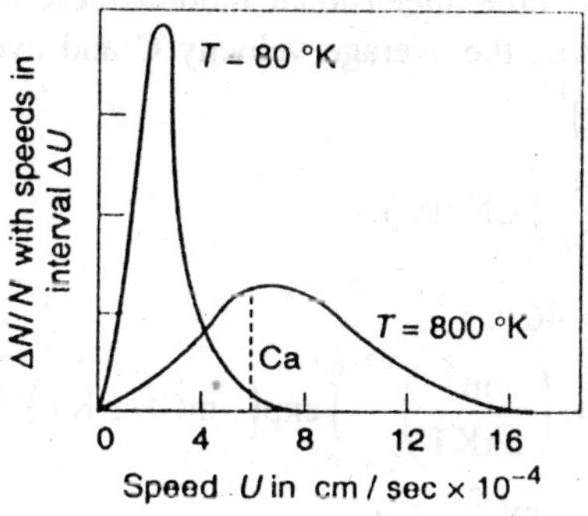

Fig. 3.9

$$C^2 = U_x^2 + U_y^2 \quad ...(45)$$

$$\frac{DN(c)}{N} = \frac{m}{2\pi KT} \exp\left(-C^2/2KT\right) Cdc \quad ...(46)$$

Extending the arguments to three dimensions

$$c^2 = U_x^2 + U_y^2 + U_z^2 \; \frac{dN(C)}{dN} \quad ...(47)$$

$$= 4\pi\left(\frac{m}{2\pi KT}\right)^{3/2} \exp\left(-mC^2/2KT\right)C^2 dc \quad ...(48)$$

gives the plot of equation 48 at different temperatures.

Most Probable Velocity

The value of C, for which $\frac{dN(C)}{dN}$ is maximum, is called the most probable velocity C_p. It is given by

$$\frac{-dC_p}{dC} = \frac{-m}{2\pi KT} [\exp(-mC^2p/2KT)\; 2C_p\; (C^2{}_p)$$

$$+ [\exp(-mC^2{}_p/2\;KT)]\;(2C_p) \quad ...(49)$$

Since $\frac{-dC_p}{dC} = 0$

$$C_p^2 = \frac{2KT}{m} \quad ...(50)$$

Average Velocities

In this case of three-dimensional motion there are two types of average velocities viz. the average velocity C and average root mean square velocity $\left(\overline{C}\right)^{1/2}$

$$\overline{C} = \frac{1}{N}\int_0^{\alpha} CNd(C) \quad ...(51)$$

From equation 46

$$\overline{C} = 4\pi\left(\frac{m}{2\pi KT}\right)^{3/2} \int_0^{\alpha} \exp\left(-mC^2/2\;KT\right)C^3\; dC \quad ...(52)$$

Let $\frac{mC^2}{2KT} = x^2$

$$Cdc = 2KT\frac{x}{m}dx \quad ...(53)$$

or $$C^2 dc = \left(\frac{2KT}{m}\right)^2 x^3\, dx$$

$$\overline{C} = 4\pi\left(\frac{m}{2\pi KT}\right)^{3/2}\left(\frac{2KT}{m}\right)^2 \int_0^{\alpha} x^3 \exp\left(-x^2\right)dx$$

since $$x^3 \exp\left(-x^2\right)dx = \frac{1}{2}$$

$$\overline{C} = 2\pi\left(\frac{m}{2\pi KT}\cdot\frac{2KT}{m}\right)^{3/2}\left(\frac{2KT}{m}\right)^{1/2} = \left(\frac{8KT}{\pi m}\right)^{1/2} \quad ...(54)$$

$$\overline{C}^2 = 4\pi\left(\frac{m}{2\pi KT}\right)^{3/2}\int_0^{\alpha} \exp\left(-mC^2/2KT\right) C^4\, dc$$

$$\text{Let} = \frac{mC^2}{2KT}X^2$$

$$Cdc = \frac{mKT}{m}Xd_x$$

From equation (54)

$$C^3 = \left(\frac{2KT}{m}\right)^{3/2} X^3$$

$$C^4\, dC = \left(\frac{2KT}{m}\right)\left(\frac{2KT}{m}\right)^{3/2} X^4 \quad ...(55)$$

$$\overline{C}^2\ 4\pi = \left(\frac{m}{2\pi KT}\right)^{3/2}\left(\frac{2KT}{m}\right)^{5/2}\int_0^{\alpha} X^4 \exp\left(-x^2\right)d_x \quad ...(56)$$

But $$X^4 \exp\left(-X^2\right)d_x = \frac{3}{8}\sqrt{\pi}$$

$$\overline{C}^4 = 4\pi \times \frac{3}{8}\sqrt{\pi}\left(\frac{m}{2\pi KT}\right)^{3/2}\left(\frac{2KT}{m}\right)^{5/2} = \frac{3KT}{m} \quad ...(57)$$

$$= \frac{3PV}{Nm} = \frac{3P}{\rho} \quad ...(58)$$

where ρ is the density of the gas and is given by $\frac{Nm}{V}$

The three characteristic velocities of a gas in three dimensions are:

Most probable velocity $C_P = \left(\frac{2\ KT}{m}\right)^{1/2}$

Average velocity $\overline{C} = \left(\frac{8\,KT}{\pi\,m}\right)^{1/2} = 1.13\,C_p$ ...(59)

Root mean square velocity $\left(\overline{C}\right)^{1/2} = \left(\frac{3\,KT}{m}\right)^{1/2} = 1.22\,C_p$

i.e., $\left(\overline{C}\right)^{1/2} : \overline{C} : C_p = 1 : 0.92 : 0.815$

The distribution of the three velocities and the effect of temperature on C are shown in Fig. 3.9b where $\frac{\Delta N}{N}$ is plotted against molecular speed.

Average Kinetic Energy

$$\overline{K.E.} = \frac{1}{2}\,\overline{m}\overline{C}^2 = \frac{m}{2N_0}\int_0^{\alpha} C^2 dm$$

$$= \frac{3}{2}\,KT = \frac{3}{2}\left(\frac{R}{N_o}\right)T \quad ...(60)$$

The average kinetic energy per molecule of an ideal gas is proportional to the absolute temperature and is independent of the kind of molecule.

It can be seen from the above that the fraction of the molecules depends on two factors.

(1) The square of the velocity

(2) $\exp\left(-\frac{1}{2}mC^2/KT\right)$

We know that $\frac{1}{2}\,mC^2 = \frac{1}{2}\,mX^2 + \frac{1}{2}\,m^2 + \frac{1}{2}mZ^2$

In the random walk model one velocity component is independent of the value of the other. Therefore, the kinetic energy of a molecule is the sum of its kinetic energy components and the magnitude of one velocity component is independent of the value of other velocity components.

Similarly,

$$\exp\left(-\frac{1}{2}\,mC^2\;KT\right) = \left(\exp -\frac{1}{2}\,\frac{mX^2}{KT}\right)$$

$$\left(\exp -\frac{1}{2}\,\frac{mY^2}{KT}\right)\left(\exp -\frac{1}{2}\,\frac{mZ^2}{KT}\right) \quad ...(61)$$

The kinetic energies associated with the velocity components must be additive but the probabilities of their individual values must be multiplicative.

DERIVATION OF THE EQUATION OF STATE OF AN IDEAL GAS

Let there be a spherical vessel of radius r (Fig. 3.10). The molecules of gas are colliding with the surface of the vessel with velocity $\overline{U}$. Any force exerted on the surface of the wall by a colliding particle is balanced by an external opposite force. The net force being then zero, no motion of wall occurs. Let us isolate a gas particle and follow it. The particle that elastically collides with the wall of the vessel rebounding at an angle ϕ, equal to its angle of approach. Since the sphere is symmetrical the particle will always be in the same plane passing through the centre of the sphere. The distance between collisions is 2r cos ϕ. If δt is the time between the collisions, then

$$\delta t = \frac{c \cos \phi}{\overline{U}} \quad ...(62)$$

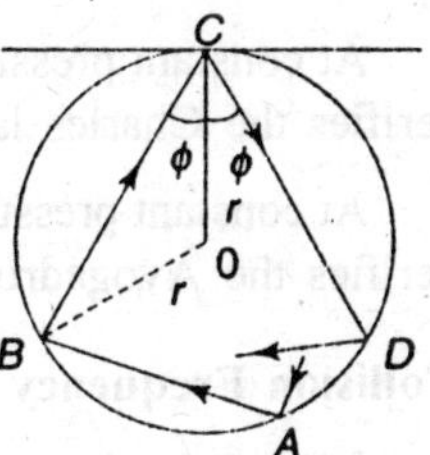

Fig. 3.10

The number of collision per unit time is

$$\frac{1}{\delta t} = \frac{\overline{U}}{2n \cos\phi} \quad ...(63)$$

The component of the velocity along the direction perpendicular in the spherical surface is U cos ϕ.

The component of the momentum of the particle of mass m before collision is –mU cos ϕ and + mU cos ϕ after the collision. The change in the momentum = 2mU cos ϕ

The force exerted by the particle on the wall

$$= \frac{2, U \cos \phi \times \overline{U}}{2r \cos \phi} = \frac{m\overline{U}^2}{r} \quad ...(64)$$

The outward force due to collision of A molecules

$$= \frac{A\, m\overline{U}^2}{r} \quad ...(65)$$

The area of the sphere = 4 π r^2

$$\text{Pressure} = \frac{Am\overline{U}^2}{4\pi r^2 \times r} = \frac{am\overline{U}^2}{4\pi r^3} \quad ...(66)$$

The volume V of the sphere

$$= \frac{4}{3}\pi r^3$$

$$\text{Pressure } P = \frac{Am\,\overline{U}^2}{3V} \quad ...(67)$$

or
$$PV = \frac{1}{3}\,A\,m\overline{U}^2 = \frac{2}{3}A\left(\frac{1}{2}m\overline{U}^2\right)$$

$$= \frac{2}{3}\text{A.K.E. (Kinetic energy)} \quad ...(68)$$

$$= RT \text{ for 1 mole of gas} \quad ...(69)$$

$$\text{Average kinetic energy} = \frac{3}{2}RT$$

or T is a measure of KE (Kinetic energy).

Since in a gas at a fixed temperature A, w and $\overline{U}$ are fixed PV is constant which verifies the Boyles law.

At constant pressure and temperature. V is proportional to T which verifies the Charles law.

At constant pressure and temperature, V is proportional to A, which verifies the Avogadro's law.

Collision Frequency

Let us consider the molecules hitting a plane surface.

As already explained that all directions in space are equivalent, the results will be expected independent of direction. According to Boltzmann distribution law the molecules with a volecity of w,

$$\frac{dN(u)}{N} = \left(\frac{m}{2\pi KT}\right)^{1/2}\left[\exp\left(\frac{-mu^2}{2KT}\right)\right] \quad ...(70)$$

The number dn of molecules per unit volume reaching the wall with velocities between u and u + du

$$dz(u) = N\left(\frac{m}{2\pi KT}\right)^{1/2}\left[\exp\left(\frac{-mu^2}{2KT}\right)\right]udu \quad ...(71)$$

z will give the total surface collision frequency per unit area, which can be obtained by integrating the above from zero to infinity.

$$z = N\left(\frac{KT}{2\pi m}\right)^{1/2} \quad z = N\left(\frac{KT}{2\pi m}\right)^{1/2} \qquad ...(72)$$

$$= \frac{1}{4} N\overline{V} = N\left(\frac{RT}{2\pi M}\right)^{1/2} \qquad ...(73)$$

$$= P\left(\frac{1}{2\pi m\, KT}\right)^{1/2} \qquad ...(74)$$

The mass collision frequency mz is given by

$$mz = \rho\left(\frac{KT}{2\pi m}\right)^{1/2} = \rho\left(\frac{RT}{2\pi M}\right)^{1/2} \qquad ...(75)$$

where ρ is the density of the gas.

Bimolecular Collision Frequency

Molecules undergo collision as they move through the gas. This will result in the change of direction and speed of the colloiding molecules. They will behave as hard spheres of radius r and undergo elastic collision so that the initial and final kinetic energies do not change.

Collision Diameter

The molecules are treated as rigid noninteracting spheres. A moving molecule will collide with other molecule whose centres come within a distance d_{12}. Thus, the molecular collision diameter d_{12} is the minimum distance of approach of two molecules colliding with zero initial kinetic energy. If r_1 and r_2 are the radii of the two molecules, then d_{12} is equal to the sum of r_1 and r_2.

The quantity πd_{12} is called scattering cross-section of the molecules and is represented by σ_{12}.

$$mz = \rho\left(\frac{KT}{2\pi m}\right)^{1/2} = \rho\left(\frac{RT}{2\pi M}\right)^{1/2} \qquad ...(76)$$

where μ is the viscosity coefficient.

Number of Collisions

The molecule A is approaching towards molecule B (Fig. 3.11 a) and shall collide with it whenever the distance between their centres become d_{12}. Let $\overline{U}$ be the average relative speed and n the number of molecules per ml. In unit time the sphere of influence of A will sweep $\pi n d^2_{12}\overline{U}$. The number of collision experienced by A will be $\pi n d^2_{12}\overline{U}$ per second.

Figure (3.11b) gives the relative velocities of two molecules just before and after the collision or just grazing collisions. The average may be taken as collision at 90°. Therefore, the relative velocity will be $2^{1/2}\overline{U}$.

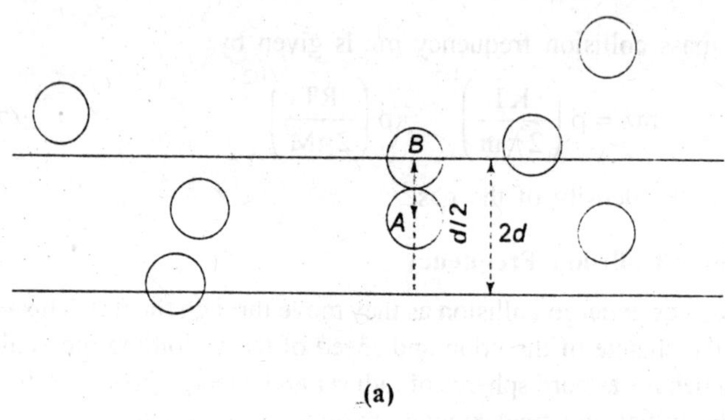

(a)

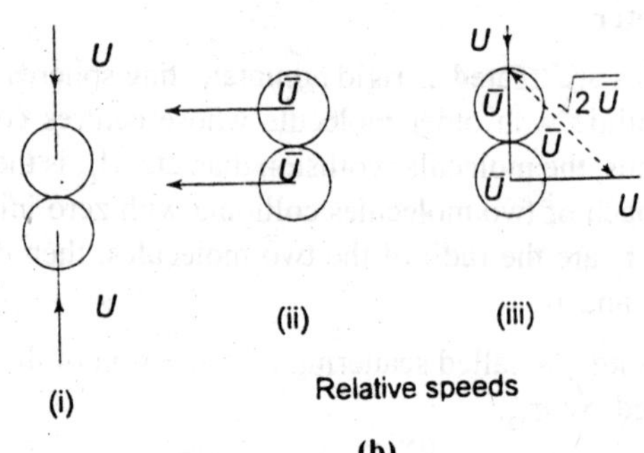

(b)

Fig. 3.11

The number of collisions z_1 will be given by

$$z_1 = 2^{1/2}\,\pi\, n d_{12}^2 \overline{U} \qquad \text{...(77)}$$

$$\text{but} \qquad = \overline{U}\left(\frac{8\,KT}{m}\right)^{1/2} \text{ and } n = \frac{P}{KT}$$

$$z_1 = 2^{1/2}\, d_{12}^2\, \pi\left(\frac{8\,KT}{m}\right)^{1/2} \quad n = \frac{P}{KT}$$

$$= 4\, d_{12}^2\, \pi \left(\frac{\pi}{mKT}\right)^{1/2} \quad ...(78)$$

From Eq. (77)

$$= \frac{m}{2(2)^{1/2}\, d_{12}^2}\left(\frac{8KT}{m}\right)^{1/2} = \left(\frac{8KT}{}\right)^{1/2} \frac{1}{d_{12}^2}$$

$$z_1 = 4P \quad ...(79)$$

Similarly, when the two molecules are similar

$$Z_{11} = \frac{1}{2}(2)^{1/2}\, \pi n^2\, \overline{V} \quad ...(80)$$

In case the two molecules are different, then the number of collisions

$$= 2(2)^{1/2}\, \pi d_{12}^2\, \overline{U}\, n_1 n_2$$

$$Z_{12} = 2(2)^{1/2}\, \pi d_{12}^2 \left(\frac{8KT}{\pi\, mr}\right) n_1 n_2$$

$$= \left(\frac{n_1}{V}\right)\left(\frac{n_2}{V}\right) 4\pi d_{12}^2 \left(\frac{KT}{2\pi m_r}\right)^{1/2} \quad ...(81)$$

where reduced mass $m_r = \frac{m_1 m_2}{m_1 + m_2}$ and $d_{12} = \frac{d_1 d_2}{2}$

$$Z_{12} = \frac{4\pi d_{12}^2\, p_1 p_2}{KT(2\pi m_r KT)^{1/2}} \quad ...(82)$$

where p_1 and p_2 are the respective pressures.

Mean Time Interval

The mean time interval between collisions

$$\tau = \frac{1}{Z} = \frac{(\pi m KT)^{1/2}}{4\pi d_{12}^2\, P} \quad ...(83)$$

Mean Free Path

The mean free path between collisions

$$\lambda = \frac{\overline{U}}{Z_1} = \frac{1}{2^{1/2}\, \pi n d_{12}^2} = \frac{\sqrt{2}\, KT}{2\pi d_{12}^2\, P} \quad ...(84)$$

The free path depends upon the number, density and size of the molecules.

If the molal volume of the gas is 50 ml/g mole then the diameter of a spherical molecule will be 5×10^{-8} cm. At NTP n is 2.7×10^{19} molecules/cm^3 and is 3×10^{-6} cm. A molecule of hydrogen will experience 10^{11} collisions/sec and the ternary collision frequency will be 10^{33} collisions per sec. cm^3. Table 3.5 gives mean free path and other values for some gases.

Table 3.5

Gas	*Molecular diameter*	*Average velocity at 20°cm/sec*	*Mean free path cm at 75 cm Hg 20°C*	*Collision frequency*
AT	2.88×10^{-8}	395×10^2	9.88×10^{-6}	4000×10^4
H_2	2.40×10^{-18}	1755×10^2	17.44×10^{-6}	10060×10^4
O_2	3.61×10^{-8}	-440×10^2	9.93×10^{-6}	4430×10^4
N_2	3.15×10^{-8}	471×10^2	9.29×10^{-6}	5070×10^4
CO	3.19×10^{-8}	471×10^2	9.23×10^{-6}	5700×10^4
CO_2	3.34×10^{-8}	376×10^2	6.15×10^{-6}	6120×10^4
NH_3	2.97×10^{-8}	604×10^2	9.6×10^{-6}	9150×10^4

EFFUSION

The process by which a gas passes from a higher pressure to a lower pressure through a minute orifice so small, that the flow is in terms of molecules rather than bulk, is known as effusion. The distribution of velocities of gas molecules remaining in the vessel is not affected by effusion.

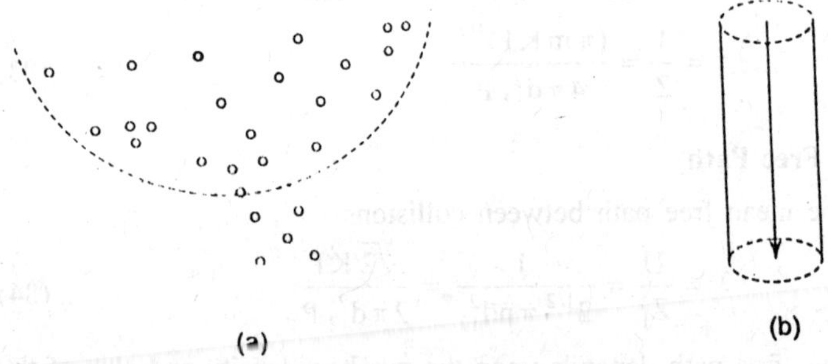

Fig. 3.12

Let da be the area of the orifice and n be the number of molecules per c.c., moving with mean velocity $\overline{U}$ (Fig. 3.12a) from any one direction will be those contained in a cylinder of base da and height $\overline{U}$ (Fig. 3.12b). The number of molecules striking the orifice would be 1/2 n$\overline{U}$ da or 1/2 n$\overline{U}$ per unit area, since only half of all the molecules are moving towards the opening.

Then using spherical polar coordinates, it can be shown that the total number of molecules striking from all directions will be given by

$$\frac{dn}{dt} = \int_0^{\pi/2}\int_0^{2\pi} \frac{1}{4\pi} n\,\overline{U} \cos\phi \sin\theta \, d\phi \, d\theta \qquad ...(85)$$

$$= \frac{1}{4} n\overline{U} = \frac{1}{4} \rho\overline{U} \qquad ...(86)$$

where ρ is the density of gas.

The average kinetic energy $= \frac{1}{2} m\overline{U}^2$

But $\left(\overline{U}^2\right)^{1/2} = \left(\frac{3RT}{M}\right)^{1/2} = 1.58 \times 10^4 \left(\frac{T}{M}\right)^{1/2}$ cm/sec.

Mass Effusion Rate

The mass effusion rate at constant temperature and pressure varies inversely as the square root of the molecular weight (Graham's law)

$$\frac{dm}{dt} = \frac{1}{4} \rho\overline{V} = \rho\left(\frac{RT}{2M}\right)^{1/2} = \frac{\alpha}{4} P\left(\frac{3M}{RT}\right) \qquad ...(87)$$

where $\alpha = \frac{\overline{V}}{\left(\overline{V}\right)^{1/2}}$

α has a value approximately equal to 0.92. Knudsen was successful in preparing thin holes for gaseous diffusion and the molecular weights of gases were found out at low pressures.

At standard temperature and pressure $\frac{n}{V}$ is 2.6870 × 10^{19} molecules per cm^3.

This is known as Loschmidt number

$$m = \alpha = \frac{1.755 \times 10^{24}}{\sqrt{M}} \text{ cm}^{-2} \text{ per sec.} \qquad (4.88)$$

m = collision frequency

Knudsen based on the above, developed a method for finding out the vapour pressure of slightly volatile substances.

$$p = w\left(\frac{2\pi RT}{M}\right)^{1/2}$$ where w is the sample weight loss per unit twice.

BAROMETRIC FORMULAE

Boltzmann derived the distribution formula which relates the particle density in different regions of an isothermal ideal gas to the potential energies of those regions.

$$(d)_2 = (d)_1 \exp\left(-\frac{U(2) - U(1)}{KT}\right) \quad ...(89)$$

where d(1) and d(2) are the densities of particles in the region of potential energies U(1) and U(2) and K, the Boltzmann constant.

The potential energy of a molecule in a uniform gravitation field

$$U = mgh \quad ...(90)$$

where m is the mass, h is the height above ground level and g = 980.665 dynes/g, is the Gaseous State due to gravity.

Since $P = \frac{n}{V} KT$ ...(91)

$$P(h) = P(O) \exp -\left(\frac{wg}{RT}\right) h = P_0 \exp \frac{-E_p}{RT}$$

where E_p is molar energy ...(92)

This is known as barometric formula and Fig. 3.13 gives the curves showing by and large the above expression is followed.

The above expression by slight modification can be applied to particles in liquids also. By modifying Eq 90.

$$U = \frac{m}{d_m}(d_m - d_l)gh \quad ...(93)$$

where d_m and d_l are the densities of particles and the liquid respectively.

Svedberg using above equation for a gold sol calculated the value of K. Since the value of R is accurately known, he obtained the value for Avogadro's number as 6.2×10^{23} which was quite near the critical value of 6 0028 × 1023.

Equation 92 depends on the molecular weight of the gas. In a gravitational field a change is expected in the composition of mixture of gases. The same behaviour is expected in the case of liquids. This principle is utilized in separating colloidal solutions in a centrifuge which increases the value of g.

HEAT CAPACITIES: TRANSLATIONAL, ROTATIONAL AND VIBRATIONAL DEGREES OF FREEDOM

So far we have shown how by using a simple molecular model we are able to derive a theoretical equation that is consistent with the empirical gas law and also we find that we can interpret the absolute temperature as a measure of the average molecular kinetic energy of translation. Let us pursue further and see if we can identify another macroscopic property, heat capacity, with the kinetic energy calculated by means of kinetic theory.

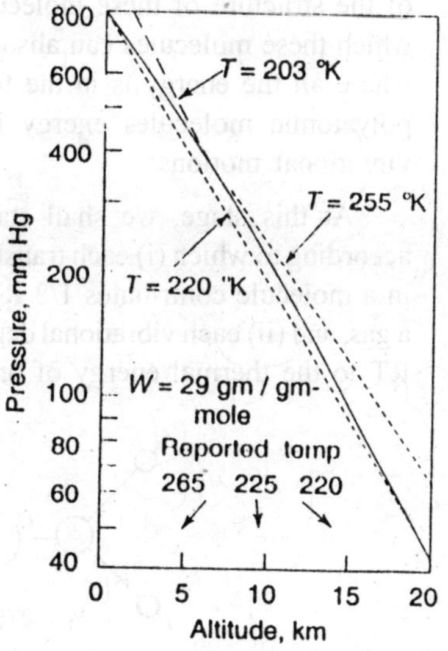

Fig. 3.13

From the fundamental equation of kinetic theory, the average translational energy for 1 mole of gas is

$$KE = \frac{1}{2} m N_0 \bar{u}^2 = \frac{3}{2} RT \quad ...(94)$$

If all the thermal energy of a gas is in the form of translational energy we can write

$$E_M = \frac{3}{2} RT \quad ...(95)$$

where the symbol E_M stands for the molar thermal energy. The heat capacity of any substance is defined as the quantity of heat required to raise the temperature by one degree and is equal to the increase in thermal energy of the system for this change of temperature. Hence, the heat capacity per mole, called the molar heat capacity $\left(\overline{C}_V\right)$ at constant volume is

$$\overline{C}_V = \frac{dE_M}{dT} = \frac{3}{2} R \quad ...(96)$$

where kinetic energy remains the same as before and after the collision. Such collisions are called elastic collisions. Since R has a value of 1.987 cal/deg. mole this equation predicts the heat capacity of a gas to be approximately 3.0 cal/deg. mole and to be independent of temperature.

The prediction of the kinetic model is accurate for monoatomic gases but the observed heat capacities of diatomic molecules are quite different. It is reasonable to assume that the discrepancy for diatomic and polyatomic molecules is connected with the increased complexity of the structure of these molecules and in particular with the ways in which these molecules can absorb energy. Unlike in monoatomic gases where all the energy is in the form of kinetic energy, in diatomic and polyatomic molecules energy is also associated with rotational and vibrational motions.

At this stage, we shall state the law of equipartition of energy according to which (i) each translational and rotational degree of freedom in a molecule contributes 1/2 RT to the thermal energy of one mole of a gas, and (ii) each vibrational degree of freedom in a molecule contributes RT to the thermal energy of one mole of a gas.

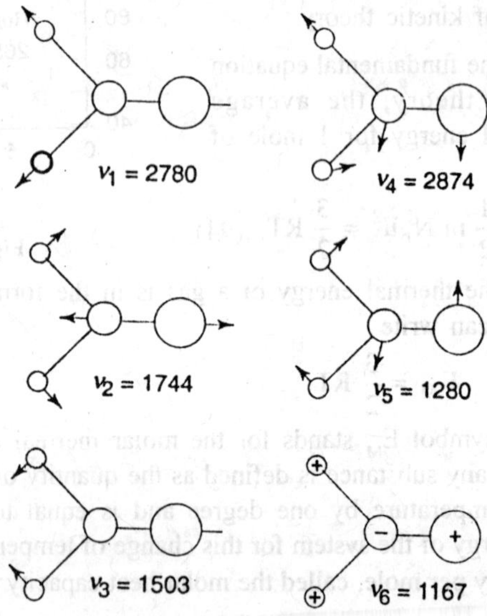

Fig. 3.14

The degrees of freedom in a molecule are given by the number of coordinates required to locate all the mass points (atoms) in a molecule. If the molecule is linear, it has two rotational degrees of freedom; for a non-linear molecule, there are three rotational degrees of freedom. The remaining degrees of freedom, that is 3N-5 for linear and 3N-6 for non-linear molecules are the vibrational degrees of freedom. Table 3.6 lists the degrees of freedom for several molecules. Fig. 3.14 gives a pictorial representation of various modes of motion for a typical polyatomic molecule.

Table 3.6 : Degrees of Freedom in Gaseous Molecules

		trans	***rot***	***vib***	***Total***
Monoatomic	He	3	0	0	3
Diatomic	N_2	3	2	1	6
Triatomic:					
Linear	CO_2	3	2	4	9
Non-linear	H_2O	3	3	3	9

In a monoatomic molecule, $E_m = 3/2\ RT$ in agreement with the simple model. For a diatomic molecule, there are three translational, two rotational (because molecule is linear) and one vibrational degrees of freedom making a total of 6. The thermal energy per mole would, therefore, be,

$$E_m = \left(\frac{3}{2}RT\right)_{trans} + \left(\frac{2}{2}RT\right)_{rot} + (1RT)_{vib}$$

and $$\bar{C}_V = \frac{3}{2}R + R + R = \frac{7}{2}R = 7 \text{ cal deg}^{-1} \text{ mole}^{-1}$$

The observed values of C_v for diatomic molecules deviate greatly from the predicted value. The fact that the value of 5 cal deg^{-1} (which is close to 5/2R) is most common for simple diatomic molecules suggests the possibility that heat capacity in these molecules is coming from translation and rotation and that vibrational motion is not contributing to the thermal energy at this temperature.

A proper explanation of the heat capacities of diatomic and polyatomic molecules is given by the quantum theory, according to which translational, rotational and vibrational energy changes can take place only in finite increments or quanta. At low temperatures, it is only translational energy which contributes to the heat capacity; as the temperature is raised, there will be rotational contributions followed by vibrational contributions. For most diatomic molecules at room temperature, all the translational and rotational degrees of freedom are fully 'active' while the vibrational degree of freedom becomes 'active' only at higher temperatures. This is why the heat capacity of all diatomic gases approaches a value of 7 cal deg^{-1} $mole^{-1}$ at high temperature.

From the equipartition theorem, we can predict the high temperature heat capacity of monoatomic solids. As the atoms in a solid can undergo only vibrational motion, therefore, each atom has three vibrational degrees of freedom. Thus, the thermal energy for a mole of solid is,

$$E_K = 3\ RT$$

and the molar heat capacity is $\overline{C}_V = 3\ R$, *i.e.*, 6 cal deg^{-1} $mole^{-1}$. This equation agrees with the empirical results of Dulong and Petit according to which the heat capacity of one gram atom of a solid element is approximately 6.3 cal deg^{-1}.

DULONC AND PETITS LAW

In 1819, Dulong and Petit observed that product of the atomic weight and the specific heat is 6.4

$$C_\upsilon\ (\text{Cal/g deg}) \times M\ (\text{g/gm atm}) = 6.4\ (\text{cal/deg g atom}) \qquad ...(97)$$

The heat required to raise one mole of material by one degree centigrade is called heat capacity of the substance. The above law has been utilized to find the correct atomic weight of the substances. If the gas is heated at constant pressure, an additional work equal to $\Delta(PV)$ has to be done. Then the specific heat at constant pressure

$$C_p = C_\upsilon + \Delta(PV)$$

$$= C_\upsilon + R \text{ for ideal gas} \qquad ...(98)$$

$$= \frac{5}{2} R$$

or at constant pressure $\Delta(PV) = PV_2 - PV_1$

$$= RT_2 - RT_1 = R\ \Delta\ T$$

$$\frac{(PV)}{T} = R$$

$$\frac{C_p}{C_\upsilon} = \frac{5}{3} = 1.67 \qquad ...(99)$$

DEVIATION OF GASES FROM IDEAL BEHAVIOUR

Compressibility

The PVT behaviour of gases has so far been presumed to follow the ideal gas equation, PV = nRT. This elegantly simple equation of state is at least approximately obeyed by almost all gases. Such behaviour is said to be ideal or perfect. When measurements are carried out at high pressures or low temperatures, no real gas is found to obey the ideal gas equation. A convenient way of showing the deviation of real gases from ideal behaviour is to write the ideal gas equation in the form, $PV = \beta nRT$ where β is unity or,

$$\beta = \frac{PV}{nRT} = 1 \text{ for an ideal gas} \quad ...(100)$$

The quantity $\frac{PV}{nRT}$ is called the compressibility factor; 13= 1 under all conditions for an ideal gas and the departure of a real gas from ideality is then measured by the deviation of the compressibility factor from unit.

$$\beta = \frac{1}{V}\left(\frac{\partial v}{\partial p}\right)_{t,n} = \frac{1}{V}\left(\frac{\partial v}{\partial p}\right)_{t} \quad ...(101)$$

But $$V = \frac{m}{W}\frac{RT}{P} \quad ...(102)$$

Thus, for an ideal gas, the isothermal compressibility coefficient is a function of its pressure only.

The deviation from ideality depends on temperature and pressure.

The quantity often exhibits both positive and negative deviations from unity.

Fig. 3.15 and 3.16(a) show some examples of the non-ideal behaviour of real gases. We see that even at 1 atm., pressure there are small deviations from ideality for all gases at any temperature. The extent of deviation at any given temperature and pressure, of course, depends on the nature of the gas, and approaches

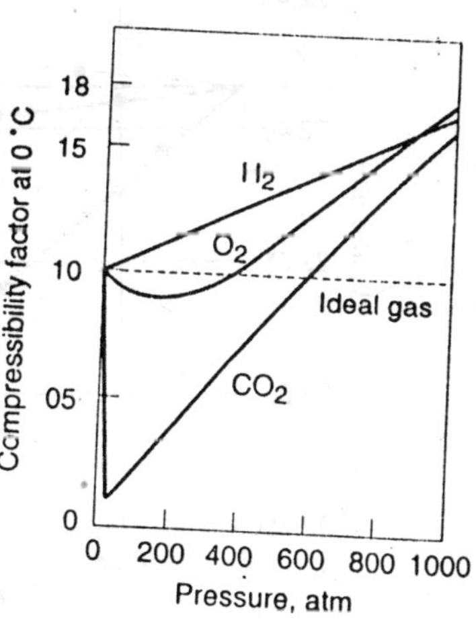

Fig. 3.15

unity only in the vicinity of zero pressure which means that PV = nRT is only a limiting law which applies exactly only at zero pressure. However, at finite but low pressure, it provides an useful approximation. The molar volume of an ideal gas (22.414 litres at STP) is not a measurable quantity but is obtained by plotting PV per mole of gases as function

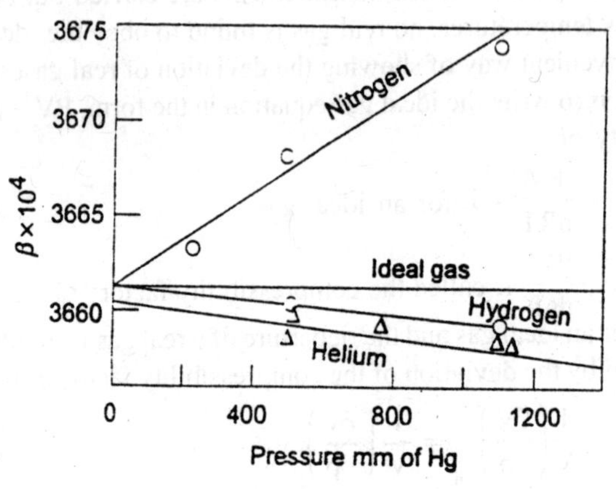

(a)

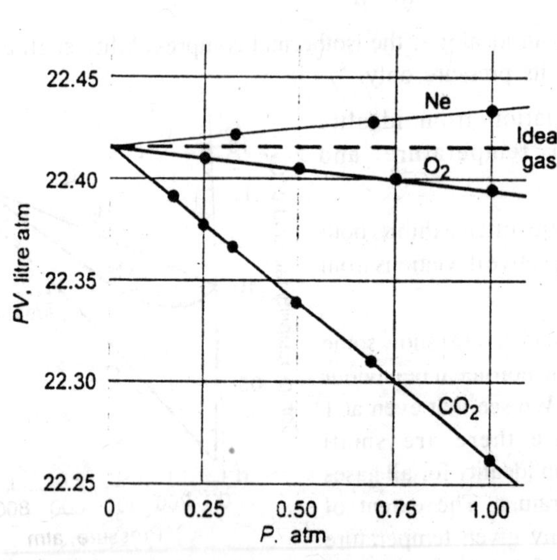

(b)

Fig. 3.16

of pressure and extrapolating the curves to P = 0 [Fig. 3.16(b)]. All extrapolated curves intersect the PV axis 22.414 since all gases behave ideally in the limit of zero pressure.

Non-Ideal Gases Non-Ideal Behaviour

In 1847 Regnault observed that the real gases do not always behave as ideal gases.

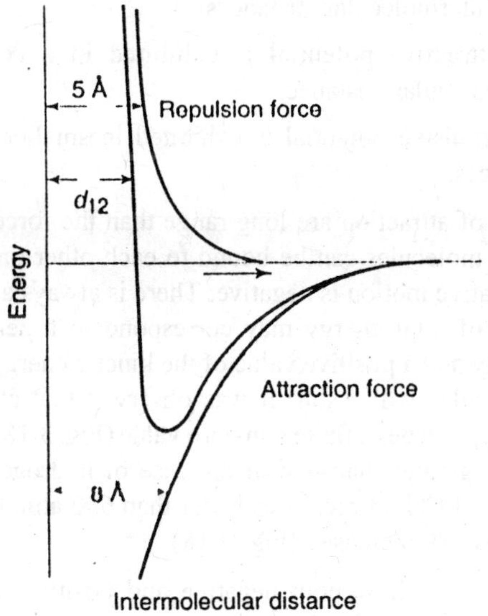

Fig. 3.17

The real gas deviate from ideality both positively and negatively. This suggests that the size of the molecule must be taken into consideration although there are difficulties in measuring it. In case we attempt to measure the size of a small sponge ball by means of a micrometer, the reading will depend on the amount of deformation produced in the ball. The correct size of the ball will be one measured without deformation. The same may be done by measuring the potential energy function of the interaction between the ball and the micrometer jaw. To resist the deformation the force must be positive and hence, repulsive. The negative deviation from ideality will be due to inherent force of attraction between the molecules otherwise the condensation of vapour to liquid cannot be explained. The compressibility of the liquid is very small, which shows that the attraction is rather limited with further pressure on the liquid, the size effect of the molecules begins to play a predominant part.

The intermolecular potential energy between the molecule is shown in the curve (Fig. 3.17). It is true only for molecule whose interaction is independent of the shape. It pertains to. spherical molecules or atoms.

The value zero energy is of the interaction at infinite distances. The following facts can be observed from the curve.

(1) The interaction potential approaches to zero for sufficiently large intermolecular distances.

(2) The attractive potential is exhibited in a certain range of intermolecular distance.

(3) The repulsive potential is exhibited in smaller intermolecular distances.

The force of attraction are long range than the forces of repulsion. Therefore, the molecules can be bound to each other only if their total energy of a relative motion is negative. There is always a possibility that a given value of total energy may correspond to a zero value of the potential energy and a positive value of the kinetic energy. In such cases unbound molecules will result. It was observed that at low values of pressure, PV approaches a finite non-zero value (Fig. 3.18). The deviation is positive and greater than one in the case of hydrogen but less than one in the case of CO_2 at pressures lower than 600 atm. but greater than one as the pressure increases (Fig. 3.18).

The real gases show both negative and positive deviations from ideal gas.

At high pressures all gases go beyond the ideal gas. At low pressures

$$PV = RT + b'P = RT(1 + bP) \quad \text{...(103)}$$

where b and b' are characteristic of gases and are functions of T.

At high pressures viral equation is followed

$$PV = R(T) + \frac{B(T)}{V} + \frac{C(T)}{V^2} + \frac{D(T)}{V^3} \quad \text{...(104)}$$

where R(T), B(T), C(T) and D(T) are known as viral co-efficients and are again functions of the nature of the gas and the temperature.

Why do gases deviate from ideality? While deriving the ideal gas equation, we made a number of assumptions. The fact that this equation is not obeyed means that some of the assumptions may not be valid. Let us re-examine two of the assumptions: (i) that the volume of the molecules is negligible compared to the empty space between them and (ii) that there are no attractive forces between the molecules.

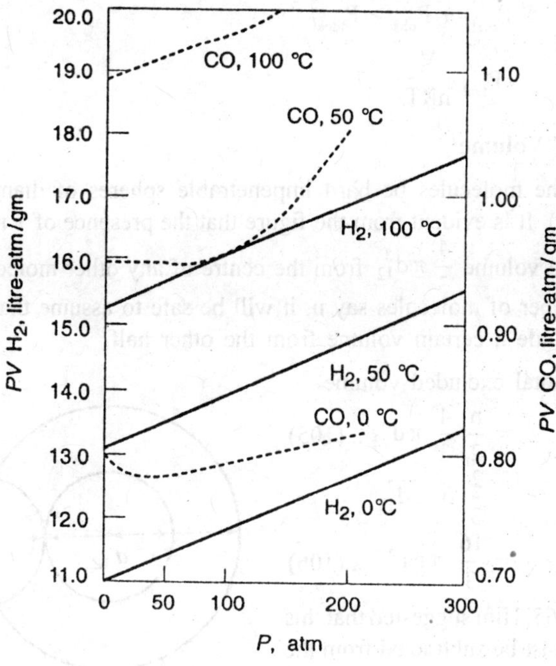

Fig. 3.18

We know that the volume of a gas can be reduced by applying pressure or by cooling the gas until the gas condenses into a liquid or solid (with finite volume). This implies that molecules in the gas must also occupy some volume, which is probably of the same order as the volume occupied by the same molecules in the solid state (note that the compressibility of solids is very small). Under normal conditions of temperature and pressure, the volume of molecules is just about 0.1 per cent of the total volume of the gas. At very high pressures (say 100 atm) or at very low temperatures, the total volume of the gas decreases appreciably (while the actual volume of molecules remains the same); under these conditions the volume of the molecules can no longer be neglected.

The pressure exerted by an ideal gas at a given temperature in a container of volume V is $P_{ideal} = RT/nV$. In real gases, the space available for the motion of molecules will be somewhat less than the volume of the container (since gas molecules occupy some volume). Because of this the observed pressure will be higher than that calculated by ideal gas law

$$P_{real} \text{ or } P_{obs} > P_{ideal},$$

or
$$P_{real} \frac{V}{nRT} > 1$$

Excluded Volume

Let the molecules be hard impenetrable spheres of diameter d_{12} (Fig. 3.19). It is evident from the figure that the presence of 1 molecule excludes a volume $\frac{4}{3}\pi d_{12}^3$ from the centre of any other molecule. For large number of molecules say n, it will be safe to assume that half of them exclude a certain volume from the other half.

The total excluded volume

$$= \frac{n}{2}\frac{4}{3}\pi d_{12}^3 \quad ...(105)$$

$$= \frac{2}{3} n \pi d_{12}^3$$

$$= \frac{16}{3}\pi n r^3 \quad ...(106)$$

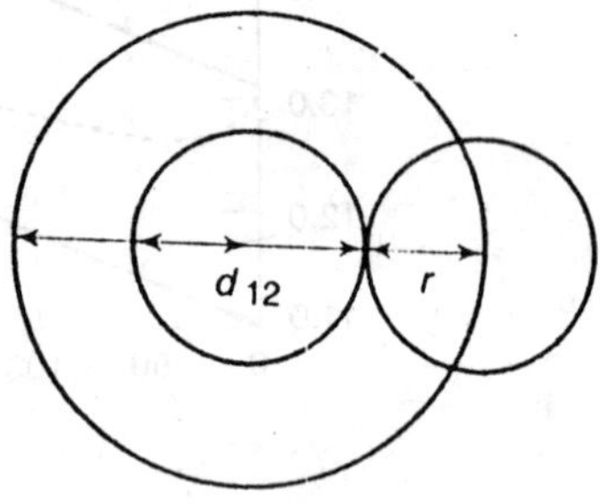

Fig. 3.19

In 1865, Him suggested that this volume must be subtracted from the volume of the gas in Boyles law. For a particular gas this is constant and can be represented by b as given by Van der Waal.

$$b = 4N (4/3 \pi r^3) \quad ...(107)$$

To obtain the correct expression the real volume (V-b) should be used,

$$P(V - b) = nRT \quad ...(108)$$

Forces of Attraction

The assumption that there are no intermolecular forces between gas molecules is a gross oversimplification. The very fact that gases can be condensed to liquids and solids indicate that there are attractive forces acting between molecules; the attractive forces become large when molecules are crowded together as in liquids and solids. At low pressures even real gases behave as ideal gases.

Joule-Thomson observed that ATP appreciable pressures, the forces of attraction and repulsion, exist even at low pressures.

A significant feature shown by the gases (Fig. 3.22) is an abrupt increase in pressure as the volume decreases to somewhat zero value which corresponds to the volume of the liquid.

$$P(V - Vg) = RT \quad \text{...(109)}$$

where V_g is the co-volume of the gas represented by the effective volume of the liquid. This only satisfies the negative variations.

In Figs 3.20(a) and 3.20(b), the two molecules are kept at a distance d_{12}. There are two forces of attraction between the proton of 1 and electron of 2 and vice versa and two forces of repulsion between the electrons and the two protons. It can be easily seen that there would be greater force of attraction between the two molecules as shown in Fig. 3.20b than in Fig. 3.20a; also it was suggested that $(P + p_1)$ should be used instead of P in ideal equation of state. This explains the positive variation.

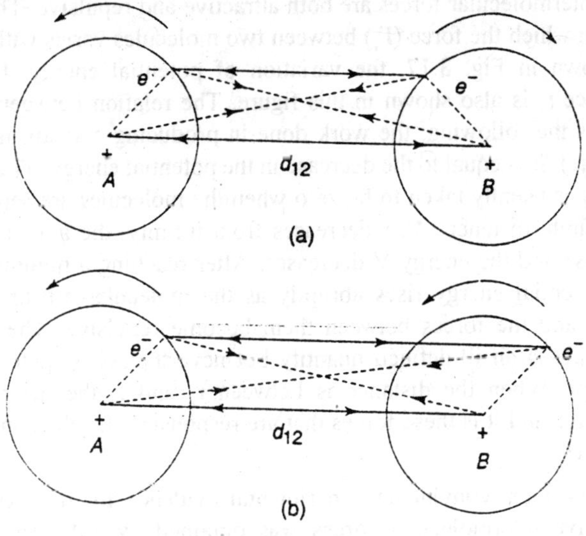

Fig. 3.20

Lennard-Jones plotted the potential energy curve (Fig. 3.17). He observed that the attractive force start working from a distance of 8Å and continue to do it till the distance between two molecules become 5Å when the repulsive force also starts working.

The potential energy is zero when the two molecules are at infinite distance from each other, but it becomes negative as they are brought near each other because of the force attraction till it reaches the minimum when the distances reach nearly 5Å and there is an abrupt increase in potential energy because of the forces of repulsion.

The forces of attraction will be proportional to the number of molecules in a given volume density.

The force $\alpha \frac{K_1}{V}$ because of one molecule

The force $\alpha \frac{K_2}{V}$ because of second molecule

Net force $= \frac{K_1K_2}{V^2}$

Van der Waal gave the symbol a for K_1K_2.

Intermolecular Forces and Joule-Thomson Effect

Intermolecular forces are both attractive and repulsive. The general way in which the force (F_r) between two molecules varies with distance is shown in Fig. 3.17, the variation of potential energy, U(r), with distance r, is also shown in this figure. The relation between F(r) and V(r) is the following; the work done in producing a small increase Δr in r, F(r) Δr is equal to the decrease in the potential energy. The potential energy is usually taken to be zero when the molecules are separated by an infinite distance. As r decreases from infinity, the attractive forces increase and the energy V decreases. After reaching a minimum value, the potential energy rises abruptly as the molecules are brought still closer and the forces between them become repulsive. The size of a molecule is an ill-defined quantity but nevertheless is quite an useful concept. When the distance is between r_e and r, the net forces are attractive and it is these forces that are responsible for the condensation of gases.

The most convincing experimental evidence for the existence of attractive intermolecular forces was obtained by J.P. Joule and W. Thomson (1852) who showed that when a compressed gas is allowed to expand in a thermally insulated apparatus, a fall in temperature is noticed. This phenomenon is called Joule-Thomson effect and was experimentally observed by allowing the real gas to adiabatically expand through a porous plug. The fall in temperature is attributed to the work done by the gas during expansion to overcome the intermolecular forces of attraction. On expansion, the distance between molecules increases and hence, the potential energy increases; this increase in potential energy must take place at the expense of the kinetic energy if the energy of the system is to be conserved. The decrease in kinetic energy is then observed as a decrease in temperature (cooling). The expansion of an ideal gas in which there are no forces of attraction would not produce the cooling effect.

$$\left(\frac{\partial T}{\partial P}\right)_H = \mu \qquad ...(110)$$

μ is known as Joule Thomson's coefficient. The three conditions are

$$\mu > 0 \text{ the gas will cool} \qquad ...(111)$$

$$\mu < 0 \text{ the gas will warm up} \qquad ...(112)$$

$$\mu = 0 \text{ no change} \qquad ...(113)$$

The temperature at which μ is zero is known as inversion temperature.

Joule-Thomson's expression is an adiabatic process so Δ H is zero

$$dH = \left(\frac{\partial H}{\partial \tau}\right)_P dT = \left(\frac{\partial H}{\partial P}\right)_T dp = 0$$

dividing by dp, $\left(\frac{\partial H}{\partial P}\right)_T = -\left(\frac{\partial H}{\partial \tau}\right)_P \left(\frac{\partial H}{\partial P}\right)_H = -c_p\mu$

where $C_P = -\left(\frac{\partial H}{\partial T}\right)_P \therefore M = \left(\frac{\partial T}{\partial P}\right)_H$

THE VAN DER WAALS EQUATION

The first successful attempt to modify the ideal gas equation in order to correct for the molecular volume and attraction is attributed to Van der Waals (1873). The Van der Waals equation of state is given as,

$$\left(P + \frac{n^2a}{V^2}\right)(V - nb) = nRT \qquad ...(114)$$

where a and b are constants; P, V, T, n and R have the same meaning as in the ideal gas equation of state. The term n^2a/V^2 is measure of the attractive forces between gas molecules and has the dimensions of pressure; this term has to be added to the measured? to get the ideal pressure. To make allowance for the finite volume of gas molecules, we subtract the term nb from the measured volume V. This nb term has the dimensions of volume and is a function of the molecular diameter. Values of a and b for a few gases are given in Table 3.7.

Table 3.8 : Van der Waals Constants

Gas	$\left(\frac{\text{litre}^2\ \text{atm}}{\text{mole}^2}\right)$	$\left(\frac{\text{litre}}{\text{mole}}\right)$
a	*b*	
He	0.0341	0.0237
H_2	0.2444	0.02661
N_2	1.390	0.03913
O_2	1.360	0.03183
CO	1.486	0.03985
CO_2	3.592	0.04267
NH_3	4.170	0.03707
H_2O	5.464	0.03049
H_2S	4.431	0.04287
Xe	4.194	0.05105
C_2H_2	4.39	0.05136
C_2H_2	4.471	0.05714
C_3H_4	2.253	0.04287
C_2H_6	5.489	0.06380
C_3H_8	8.664	0.08445
C_3H_6	8.379	0.08273
C_4H_{10}	14.470	0.1226

Now for 1 mole of a gas

$$P = \frac{RT}{V-b} - \frac{a}{V^2} \quad \text{...(114)}$$

$$\beta = \frac{PV}{RT} = \frac{1}{1-\frac{b}{V}} - \frac{a}{RTV} = \frac{1}{V}\left(\frac{\partial V}{\partial \rho}\right)_{C,m} \quad \text{...(115)}$$

$$= 1 + \frac{b}{V} + \frac{b^2}{V^2} + \ldots - \frac{a}{RTV}$$

$$= 1 + \left(b - \frac{a}{RT}\right)\frac{1}{V} + \left(\frac{b}{V}\right)^2 \qquad ...(116)$$

since b/V is small in the case of gases, higher terms may be neglected factor

Equation 116 can also be written as

$$\frac{PV}{RT} = 1 + \frac{B}{V} + \frac{C}{V^2} \qquad ...(117)$$

This is known as Viral Equation of state, also

$$\alpha \frac{1}{V}\left(\frac{\partial V}{\partial \rho}\right)_{E,m}$$

α is known as coefficient of thermal expansion.

The second and the third viral coefficients are

$$B(T) = b - \frac{a}{RT} \qquad ...(118)$$

and $\qquad C(T) = b_2$

Sometimes it is convenient to express the equation as a power series in P, by replacing $\frac{1}{V}$ by $\frac{P}{RT}$

$$\beta = 1 + \frac{P}{RT}\left(b - \frac{a}{RT}\right) + \left(\frac{4p}{RT}\right)^2 \qquad ...(119)$$

as P approaches zero, the coefficient

$$\left(\frac{\beta - 1}{p}\right) = \frac{1}{RT}\left(b - \frac{a}{RT}\right) \qquad ...(120)$$

The slope of β vs P plot should be positive at small pressures if $b > \frac{a}{RT}$ and zero if $b = \frac{a}{RT}$ (Fig. 3.21).

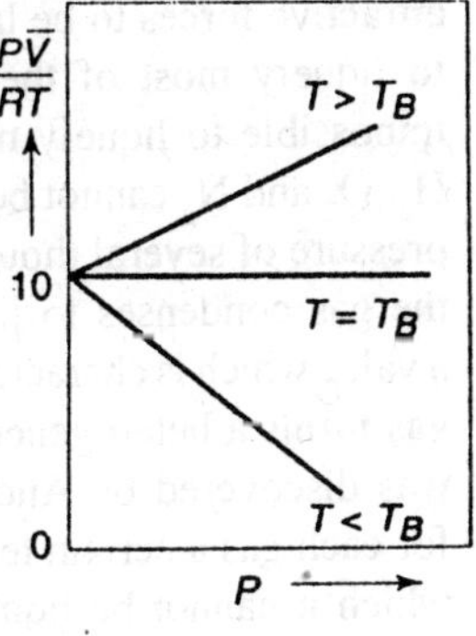

Fig. 3.21

BOYLE'S TEMPERATURE

There should be a temperature T_B for every gas, at which the gas behaves as an ideal gas. This temperature is known as Boyle's temperature, but it is not critical temperature.

$$T_B = \frac{a}{bR} \qquad ...(121)$$

In the case of the plot of β vs P, the three conditions are

(1) $T > T_B$ Positive slope

$$b > \frac{a}{RT}$$

(2) $T < T_B$ Negative slope

$$b < \frac{a}{RT}$$

(3) $T = T_B$ No change

$$b = \frac{a}{RT}$$

LIQUEFACTION OF GASES

By the middle of the 19th century a number of gases like NH_3, CO_2, H_2S and N_2O were liquefied by the combined effect of compression and cooling. Liquefaction is easy to understand from the microscopic view point since the kinetic energy of the molecules decreases on cooling. As a result of this, some of the molecules get closer to one another till they change into the liquid state. High pressure also favours liquefaction because molecules are brought sufficiently close together to permit the attractive forces to be large enough to cause condensation. It is possible to liquefy most of the gases at atmospheric pressure, but it is quite impossible to liquefy many gases at ordinary temperatures. Gases like H_2, O_2 and N_2 cannot be liquefied at laboratory temperatures even under pressure of several thousand atmospheres. At sufficiently large pressure, the gas condenses to liquid provided the temperature is maintained at a value which is characteristic of that gas. In this process the homogeneous gas forms a heterogeneous gas. The essential condition for liquefaction was discovered by Andrews (1863). He demonstrated that there exists for each gas a certain temperature, called the critical temperature, above which it cannot be liquefied, no matter how great the pressure. Young studied the pressure-volume relation of isopentene at different temperatures and his results are shown in the Fig. 3.22. The isotherm at 220°C represents the data obtained at a relatively high temperature. Its form is very close to that of a rectangular hyperbola and shows that at this temperature the gas follows Boyle's law rather closely.

Now consider the isotherm at 170°C which is below the critical point. As the vapour is compressed, the P-V curve follows AB which is roughly in accordance with Boyle's law. When point B is reached liquid isopentene is formed and this can be observed by the appearance of a meniscus between vapour phase and liquid phase. As the volume decreases further, more of the gas is transferred to the liquid phase while the pressure remains constant (note the flat portion of the curve). Finally, at C all of the gas is in the liquid phase and the curve CD is the isotherm of liquid isopentene. Since the liquids are relatively incompressible, the compression of the liquid from C to D results in little volume change and hence, the curve CD is very steep.

The flat part of the isotherm reveals an important fact that as long as both liquid and gas phases are present, the pressure of the gas in contact with the liquid must be the same, quite independent of whether a small or a large fraction of the volume is occupied by liquid. This equilibrium pressure is known as the vapour pressure of the liquid at that particular temperature. This corresponds to the heterogeneous mixture.

The specific volume V is given by

$$V_{mixture} = f_g V_g + f_l Y_l \qquad ...(122)$$

where f is a function of the gas and the liquid respectively. As isotherms are taken at successively higher temperature we find that the flat part, *i.e.*, the volume range in which the phases coexist, becomes shorter and shorter until for a particular isotherm at 187°C its length is reduced to a zero at the point P. This is the critical point a J the isotherm on which it lies is the critical isotherm. This temperature is known as the critical temperature.

The pressure at the critical point is known as the critical pressure (P_c), and the corresponding volume per mole is called the critical molar volume. It is to be noted that at the critical temperature, liquid and vapour of a substance become so similar that they can no longer be distinguished as individuals. The indexes of refraction, densities and molar volumes of the two phases become identical in the critical state. Andrews stated of this state: "If anyone asked whether it (the substance) is now in the gaseous or in the liquid state, the question does not arise I believe, admit of a positive reply."

CRITICAL CONSTANTS

For every gas there will be a critical temperature T_c, at which the horizontal region disappears and beyond which no liquefaction occurs (Fig. 3.22). Corresponding to this temperature, there will be critical volume V_c and critical pressure P_c.

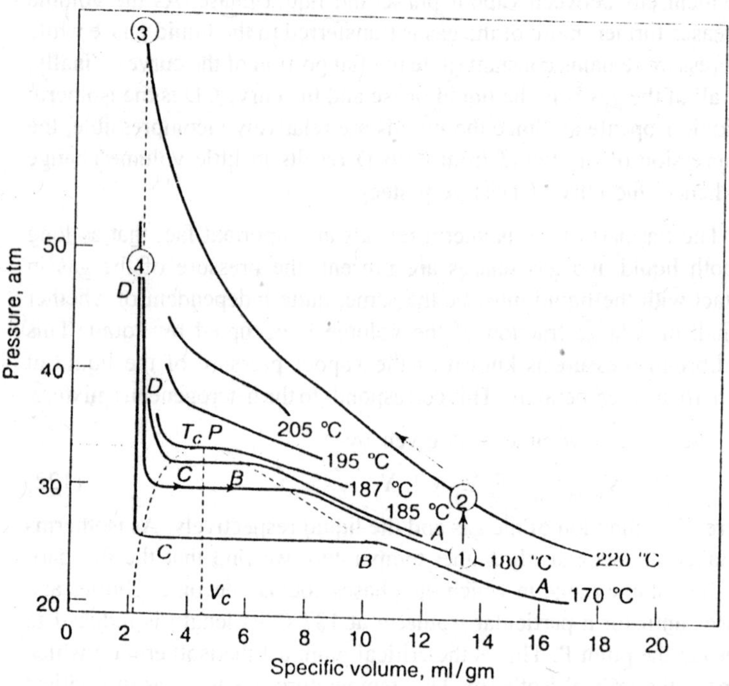

Fig. 3.22

The Van der Waals equation can be written as

$$PV^3 - (Pb + RT)\,V_2 + aV - ab = 0 \qquad \text{...(123)}$$

For a given value of a and b, the above equation will have three equal roots on T_c,

$$\left(\frac{\partial P}{\partial V}\right)_T = \frac{-RT_c}{(V_c - b)^2} + \frac{2a}{V_c^3} = 0 \qquad \text{...(124)}$$

$$\left(\frac{\partial^2 P}{\partial^2 V^2}\right)_T = \frac{2RT_c}{(V_c - b)^2} + \frac{6a}{V_c^4} = 0 \qquad \text{...(125)}$$

$$T_c = \frac{8a}{27bR},\ V_c = 3b,$$

$$P_c \;\frac{a}{27b^2},\ \frac{P_c V_c}{RT_c} = 0.375$$

or $a = 3P_c V_c^2$ and $b = \frac{V_c}{3}$ and $R = \frac{8P_c V_c}{3T_c}$

LAW OF CORRESPONDING STATES

If the variables are converted to reduced variables such that $\frac{P}{P_c} = R_R$; $\frac{V}{V_c} = V_R$ and $\frac{T}{T_c} = T_R$, the Van der Waals equation becomes

$$\left(P_R + \frac{3}{V_r}\right)(3V_r - 1) = 8T_r \qquad ...(126)$$

This contains no constant specific to the particular substance.

The law of corresponding states relates that for a group of similar gases there is one functional form of the equation of state

$$P_r = f(V_r, T_r) \qquad ...(127)$$

By plotting β as a function of reduced pressure at various reduced temperatures, it is observed that many gases behave alike in terms ul three variables (Fig. shown in example. 1).

OTHER EQUATIONS OF STATE

The Van der Waals equation provides only a reasonable correct information in the range of moderate deviations from ideality at pressures below 1 atm. It contains three arbitrary constants a, b and R.

Viral Equation

The general form of equation is

$$PV = RT + \frac{B(T)}{V} + \frac{C(T)}{V^2} + \frac{D(T)}{V^3} \qquad ...(127)$$

Equation 127 is known as viral equation, and RT, B(T), C(T)... are known as first, second, third viral coefficients.

Clausius Equation

Clausius derived an equation containing four constants

$$\left(P + \frac{a}{(V+C)^2\, T}\right)(V - b) = RT \qquad ...(128)$$

$$\frac{RTC}{P_c V_c} = \frac{8(b+C)}{3(b+2c)} \qquad ...(129)$$

$\frac{RT_c}{P_cV_c}$ will be constant only when C = bK

$$\frac{RT_c}{P_cV_c} = \frac{8(1+K)}{3(1+2K)} \qquad ...(130)$$

The experimental value of K is found to be 1.63, which gives the value of $\frac{RT_c}{P_cV_c}$ = 3.36 a value lower than the experimental value,

D. Berthelots Equation

Berthelot gave an equation which gives better results at pressures not much above 1 atm. It also involves three constants.

$$\left(P + \frac{A}{TV^2}\right)(V - B) = RT \qquad ...(131)$$

Sometimes it is used in the reduced form

$$P_rV_r = R^1\,T_r\left[1 + \frac{9}{128}\frac{R_r}{T_r}\left(1 - \frac{6}{T_r^2}\right)\right] \qquad ...(132)$$

where $R^1 = \frac{RT_c}{P_cV_c}$

Dietericis Equation

Dietericis made an allowance for the density gradient at the boundary of the gas as internal pressure dimensions to zero.

$$P(V - b) = RT_c e^{\frac{-e}{VRT}}$$

$$V_c = 2b;\ RT_c = \frac{a}{4b} \text{ and } P_c = \frac{a}{4b^2}e^{-2}$$

$$\frac{RT_c}{P_cV_c} = \frac{1}{2}e^2 = 3.695.$$

One of the best empirical equation is that proposed by Beatti and Bridgemann. This equation has five constants in addition to R and predicts accurate values (PVT) over a wide range of temperature and pressure (250 atm).

$$PV^2 = RT\left(1 - \frac{C}{VY^3}\right)\left(V + B_o - \frac{bB_o}{V}\right)$$

Benedict and Coworkers have given an equation with 8 constants. It reproduces the isotherm even in the liquid region quite well.

$$PV = RT\left(1 + \frac{bB_o}{V} - \frac{b}{V^2}\right) - \left(\frac{A_o}{V} + \frac{a}{V^2} - \frac{a}{V^5}\right) - \frac{1}{T^2}$$

$$\left(\frac{C_o}{V} - \frac{C}{V^2}\left(1 + \frac{r}{V^2}\right)eV^2\right)$$

where second viral coefficient $B = B_0 - \frac{A_o}{RT} - \frac{C_o}{RT^3}$

Table 3.9 : Values of different constants in Benedicts equation atm, litres g moles °K R = 0.08206

	A_o	a	B_o	b	$e \times 10^{-4}$
H_2	0.1975	–0.00506	0.2096	–0.04359	0.0504
N_2	1.3445	0.02617	0.05046	–0.00691	4.20
O_2	1.4911	0.02562	0.04624	0.004208	4.80
CO_2	5.0065	0.07132	0.10476	0.07235	66.00
CH_4	2.2768	0.01855	0.5587	–0.01587	12.83
C_2H_4	6.1520	0.04964	0.12156	0.03597	22.68
NH_3	2.3930	0.17031	0.03415	0.19112	476.87
C_2H_6	5.8800	0.05861	0.09400	0.01915	90.00
C_3H_8	11.9200	0.0734	0.18100	0.04293	120.00

NON-UNIFORM GASES

The values of the intensive variables of a non-uniform gas, at the same instant of time, vary from one region to another in the gas. A non-uniform gas may be regarded as composed of infinitely connected small regions of uniform gas. The state variables of a non-uniform gas are the functions of the position within the gas. Therefore, a microscopic state of the non-uniform gas requires the simultaneous specification of the previous functions at all position and times.

THERMAL CONDUCTIVITY

The passage of heat through a gas is accompanied by the changes in the temperature of each region of gas.

Two parallel plates A; and A; of equal area are kept at a fixed distance D (Fig. 3.23). The intervening region of uniform cross-sectional area is filled with the gas. They are heated to temperatures T_1 and T_2 respectively and a stationary state is maintained, *i.e.*, the temperature at every position in the gas is constant as a function of time. Now Q_z is the amount of heat per unit time flowing in the z direction provided D and $T_1 - T_2$ are not large, it can be shown that

$$Q'_z = A\, f\left(\frac{\partial T}{\partial z}\right)_{XY} \quad ...(138)$$

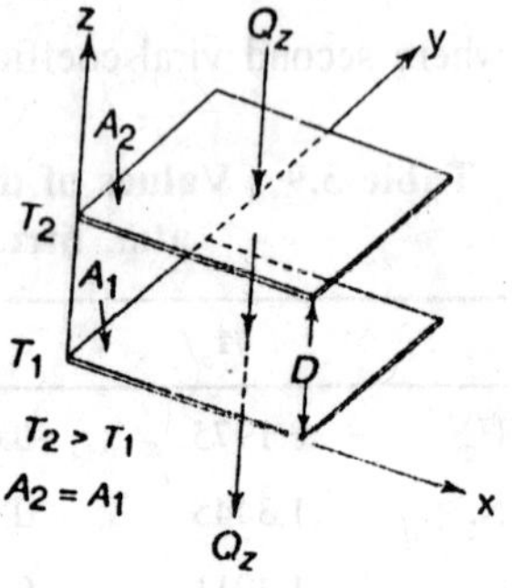

where f is a function, such that f(–x)–(x).

$$q'_z = \frac{Q'_z}{A} = \left(\frac{\partial T}{\partial z}\right)_{x,y} \quad ...(139)$$

The quantity K is called the coefficient of thermal conductivity and is independent of temperature gradient. If this is isotropic

$$q' = K\upsilon T \quad ...(140)$$

Fig. 3.23

The units are calories per ml per sec/°C.

TRANSPORT PROPERTIES

If the gases are filled in two portions of a container and the partition between the two portions is removed, then these gases will diffuse in each other and form a uniform mixture of gases. These are called the transport property of gases.

Mass Transport in Gases

P_1 and P_2 are two membranes permeable to diffusing gases (Fig. 3.24). They do not allow the gas to mix in intervening region and also prevent its mechanical motion. Though A and B gases of fixed composition is continually supplied, *i.e.*, M_1 and M_2 mixtures of constant composition are maintained at the temperature T and pressure?.

By measuring the composition before entering and after leaving, the rate of diffusion of each constituent can be measured.

It can be shown for any constituent that

$$\frac{m_k}{A} = D_1\left(\frac{C_k}{z}\right)_{XY} \quad ...(141)$$

where A is the area of the cross-section in which the constituent K is diffusing at the rate of m_kg/sec and C_K g/cm³ is the concentration in the diffusing region, D_K is called the diffusion coefficient of the Kth constituent. This equation represents the Pick's first law of diffusion.

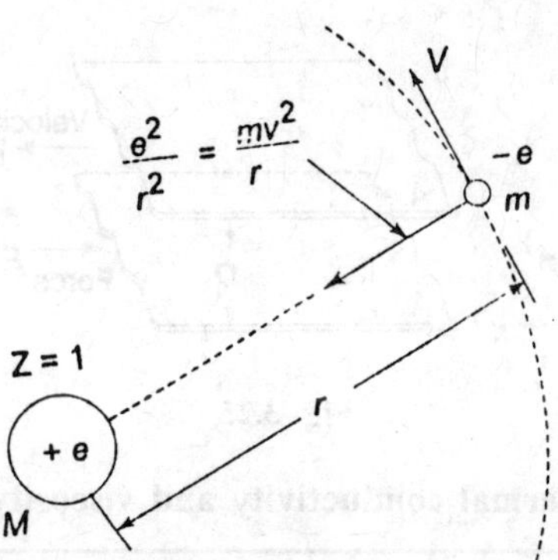

Fig. 3.24

In the case of binary mixture if the gas follows ideal gas condition, then the binary diffusion constants are the same for both the substances and also

$$\frac{1}{W_1}\left(\frac{\partial C_1}{\partial z}\right) XY + \frac{1}{W_2}\left(\frac{\partial C_2}{\partial z}\right) XY = 0 \qquad ...(142)$$

The diffusion coefficient of a substance does not depend on the substance but it depends on the medium in which diffusion occurs. No precise correlation with the molecular weight is found but substances of lower molecular weight show higher diffusibility. At low densities the diffusion coefficient of gases vary inversely as densities.

Viscous Behaviour of Gases

The gases do not offer static resistance to shear stress. On the application of very small stress they are deformed and the rate of deformation depends on their viscosity.

Let us take two plates at a distance D of gas (Fig. 3.25). The upper plate is moving with respect to the lower plate of gas which is stationary, with a velocity u cm/sec. To maintain the lower plate stationary, a force F is applied in the opposite directions. This force is opposed to one said to arise from the viscous drag of the gas, then it can be shown that

$$f = \eta \left(\frac{\partial u}{\partial z} \right)_{XY} \quad ...(143)$$

where f = F/A and η is called the coefficient of viscosity expressed as dynes sec/cm^2 which is called a poise.

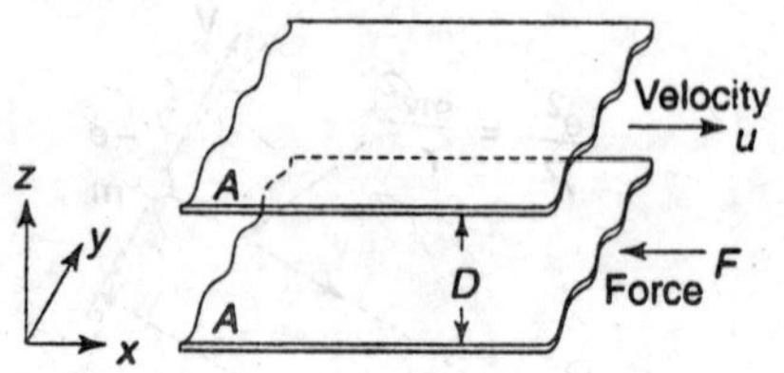

Fig. 3.25

Table 3.10 : Thermal conductivity and viscosity of some gases

x 10⁵ cm²/sec	*Cat/see cm² °C*	*× 10⁻⁶*
Diffusion coefficient	***Thermal conductivity at 0°***	***viscosity micropoises at 0°***
Acetylene	45.04	93.5
Ammonia	53.31	91.8
Butane	–33.06	109.2
–Butylene	–	74.4 at 18°
Carbon dioxide	35.72	139.0
Carbon monoxide	55.81	166.0
Ethane	44.63	84.8
Ethylene	44.63	84.8
Hydrogen	417.39	83.5
Hydrogen sulphide	31.41	116.6
Methane	74.39	102.6
Nitrogen	58.27	167.4
Oxygen	59.43	189.0
Propane	37.19	79.5 at 179°
Propylene	–	83.4 at 16.7°
Sulphur dioxide	1.950	117.0

It has no relation with the molecular weight but increases with the temperature, with the exception of very low and very high pressures, it is independent of pressure. There is a good relation between viscosity and thermal conductivity (Fig. 3.26). Similarly, there is some relation between viscosity and diffusion.

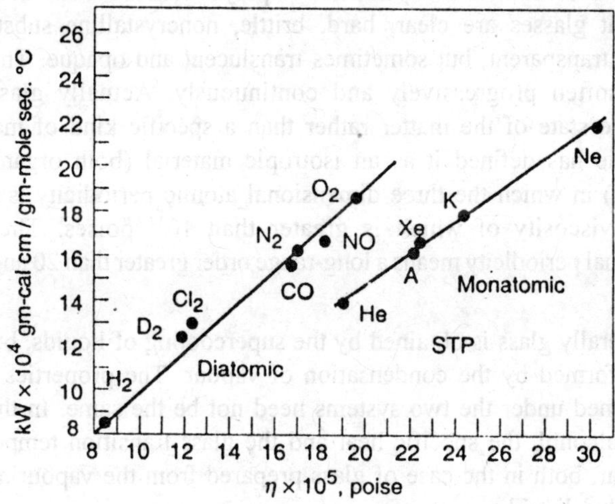

Fig. 3.26

In each of the above transport phenomenon, there is transport of either heat, fluid or force. Since these transport properties are correlated with each other, suggests that they have a common origin. It would be not wrong to assume that in each case there is transport of matter.

GLASSY STATE

Glass is a transparent, hard, rigid and brittle material. The technology of its making seems to have originated and developed' in Egypt about 3000 B.C. The interest of India in its development dates back to Indus Valley civilization as is evident by the presence of glass bangles in the excavations at Harrappa. In 1600 B.C., it was developed in India into an advanced cottage industry, manufacturing glass beads and bangles for ladies and small bottles of beautiful designs to keep essential oils and floral waters.

Definition : According to the American Society for Testing Materials (1945) (A.S.T.M.), glass was considered as a solution of inorganic products cooled to rigid condition without crystallization. Morey considers glass as an inorganic substance in a condition which is continuous with and analogous to the liquid state of that substance, but which, as a result of reversible change in viscosity during cooling, has attained so high a degree of viscosity as to be for all practical purposes rigid. Holloway defined it as a putrefied immobile liquid, which includes both inorganic and organic substances like perspex and polystyrene. Another definition states that glasses are clear, hard, brittle, noncrystalline substances, normally transparent, but sometimes translucent and opaque. These on heating soften progressively and continuously. Actually glass is a metastable state of the matter rather than a specific kind of material. Mackenzie has defined it as an isotropic material (both organic and inorganic) in which the three dimensional atomic periodicity is absent and the viscosity of which is greater than 10^{14} poises. The three dimensional periodicity means a long-range order greater than 20 angstrom units.

Generally glass is obtained by the supercooling of liquids, but they are also formed by the condensation of vapour. The properties of the glass formed under the two systems need not be the same. In the case of ethyl alcohol, the specific heat and the glass transition temperature are similar, both in the case of glass prepared from the vapour and the supercooled liquid.

INORGANIC GLASSES

Obsidion, a mineral found in nature, is a dark reddish brown, translucent glass, formed when molten rock is rapidly cooled. These inorganic glasses are the products of fusion of inorganic salts cooled to a rigid mass without crystallization. Winter observed that oxygen, sulphur, selinium, and telurium elements of group VI of the periodic table form glasses with other fifteen elements of group III, IV, and V. These group VI elements are found to form binary glasses and are self vitrifying. A discussion about these is beyond the scope of the present work and interested readers are advised to read the original work. These include various oxides, chlorides, nitrates, sulphates, thiosulphate, alums and other hydrated salts.

DEVITRIFICATION

The molten mixtures of alkali silicates, on cooling slowly, try to crystallize out into one or more of the compounds at their freezing point and the whole mass loses its glassy nature. This phenomenon is known as the devirtrification of glass. All glasses except those having very high viscosity in the molten state try to devitrify.

Porcelain

Reamer produced some polycrystalline ceramics by heating glass but it was actually Stookey who systematically studied this subject and actually produced ceramic by heating glass under controlled condition. The small devitrified glass crystal slowly grow into a strong, opaque, white mass, known as porcelain. They have a much higher mechanical strength and markedly improved electrical insulation properties than ordinary glass.

VISCOSITY TEMPERATURE CHARACTERISTICS

Glass on heating softens continuously and progressively to fluid due to continuous fall in viscosity.

Gehlhoff and Thomas found a correlation between the change in viscosity and its composition. Therefore, the relationship between viscosity and temperature is a very important parameter for the glass. Important work in this direction is carried out by Field, Royster, Herty and McCaffery. They have made a thorough study of change in viscosity with temperature and with time. Lillie has produced the most accurate results using modified Margules method. The complete viscosity temperature curve is given in Fig. 3.27. It would be of interest to define the viscosity temperature relationship in terms of certain characteristic temperature at which the glass attains a particular value for viscosity.

Certain temperatures are only of historical importance, *e.g.*, deformation temperature, cohesion temperature, flow temperature, softening point, etc., while others like working point, Mg point, annealing point, strain point and transformation point are very important in the manufacture of glasses.

Working Point

At the usual melting temperatures glass has the same viscosity as glycerine at 0°C, *i.e.*, 10^2 poise. For giving a shape to the glass, it should remain in the fluid condition over a wide range of temperature. Therefore, the glass is removed from the furnace at a viscosity of $10^4 - 10^5$, poise so that it can be given proper shape. The corresponding temperature is known as working or flow temperature.

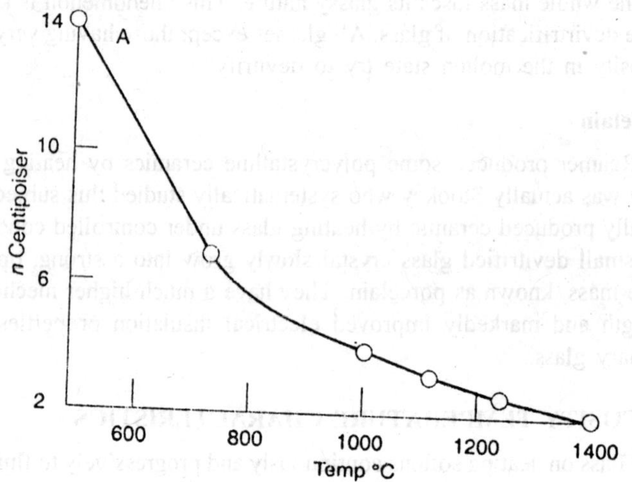

Fig. 3.27 : Complete viscosity temperature curve.

Softening Point

Littleton defines it as the temperature at which a uniform rod of 0.5 to 1.00 mm diameter and 22.9 cm long elongates under its own weight at a rate of 1 mm per minute under standard conditions. The viscosity at this temperature is $10^6 - 10^7$ poises.

Mg Point

The temperature at which the viscosity flow exactly counteracts the thermal expansion during the measurement is known as dilatometric or Mg point. According to the recommendations of the Society of Glass Technology it is the maximum point reached on the complete thermal expansion curve for the glass, *i.e.*, the point normally corresponding with the softening temperature. The viscosity at this temperatures is $10^{11} - 10^{12}$ poises. It also corresponds to the extreme upper limit of the annealing range and is denoted by A or the upper critical point designated by C_t. This temperature also depends on the thermal history of the glass.

Annealing Point

Glass on being cooled, without special heat treatment, develops unequal strained condition leading to a tendency to break. To prevent this strain, exact controlled temperature is maintained during a short interval known as annealing range. The temperature at which the internal stress in the glass is relieved within fifteen minutes, is known as annealing point. The viscosity corresponding to this temperature is $10^{13\text{-}14}$ poises.

Strain Point

Littleton defines the lower limit of the annealing point as the strain point. This is also the highest temperature from which the glass can be rapidly cooled without introducing serious internal stresses. The glass can be commercially annealed within sixteen hours. The viscosity at this temperature is $10^{14\text{-}16}$ poises.

SECOND ORDER TRANSITION

The usual change of state (solid to liquid; liquid to vapour) is called a first order transition. The free energy of the two forms, at the transformation temperature under constant pressure, is equal but there is an abrupt discontinuous change in the slope of energy vs temperature curve. There is a two-phase equilibrium, the upper and lower limits of the value at the discontinuity represents the coexisting phases.

A great many liquids on supercooling below their freezing points display a discontinuity in the temperature derivatives of energy (E), enthalpy (H), entropy (S), volume (V) whereas the quantities E, H, S, V themselves are continuous. These discontinuities are smaller below the transformation temperature than above. These are known as second order transition. It is in this region of transformation temperature that the supercooled liquids are transformed from viscous liquids into hard and brittle but still amorphous solids designated as glass, *e.g.*, liquid helium.

TRANSFORMATION TEMPERATURE

The temperature at which an abrupt change in the coefficient of expansion takes place in a glass heated at about 4°C per minute is known as the transformation point Tg. It is a characteristic temperature for a given composition. Since the molecular vibrations just begin at this temperature, it is also known as the lower critical point. Actually, it is a transition region, above which it is fluid or rubbery and below which it is glass. This is because of the interaction between various chains. According to modem point of view, it is a temperature at which configuration becomes comparable, with the rate of approach to equilibrium with the time required to make an observation. The corresponding viscosity is $10^{13} - 10^{14}$ poises.

The transition region in organic glasses is characterized by changes in derivative functions, such as specific heat and coefficient of expansion. Any molecular theory of their properties leads to molecular interpretation of the glass transition, *e.g.*, ΔC_p is close to 11.2 J.K.[7] per bead where bead is identified as the smallest unit in polymer chain capable of free rotation.

The glass transition temperature is associated with the fractional free volume having a constant value 0.025. Attempts to correlate Tg to polymer structure have focussed on either molecular cohesion or chain stiffness or both. Gibbs and coworkers predicted a second order transition in polymers at a temperature T_2 which is about 60°K below Tg and is nearly equal to the temperature at which the axtrapolated viscosity of the liquid tends to infinity. The molecular motions are found to occur in glass polymers, but in most of the cases it is not clear as to what they are. Therefore, it is safe to assume that glass transformation is associated with large-scale movements.

RELAXATION

The term relaxation expresses the establishment of a static equilibrium in a physical system. In a liquid continuous regrouping of molecular clusters appear under the influence of thermal mobility. Boltzmann's law suggests a method of working out a probability of the process. The reciprocal of this probability of regrouping gives the relaxation time, which is of the order of $10^{-4} - 10^{-6}$ sec in the case of polymers as compared to 10^{-10} sec in the ordinary liquids. Thus, polymer chains possess low mobility and higher relaxation time. The process in which equilibrium is established with time is known as relaxation process.

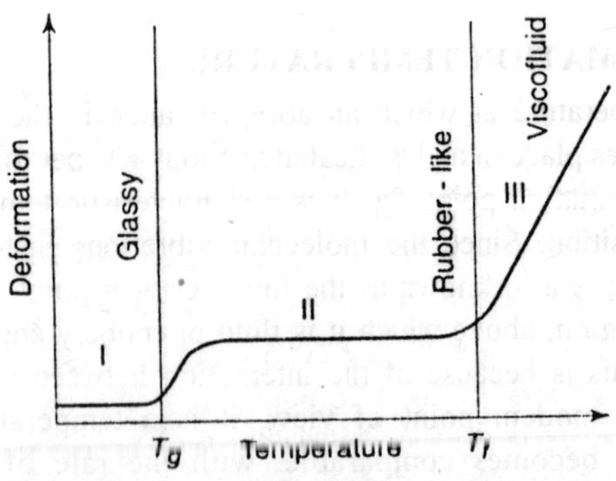

Fig. 3.28

Figure 3.28 gives a curve between temperature and deformation. This consists of three positions, namely glassy state, rubber like state and viscofluid state. The transition from the glassy state to rubber-like state involves a change in the nature of super molecular structure but the phase remains the same as there is no heat effect. The difference between glassy state and the rubber-like state is in the mobility of their macromolecules and super molecular structure. Thus, glass transition is not a phase transition.

MECHANISM OF GLASS TRANSITION

The glass transition depends upon the ratio between the intra- and intermolecular interaction energy and the thermal energy of the motion of their units. The energy of intermolecular attraction does not depend on temperature, but the thermal energy (KT) decreases sharply with decreasing temperature and at a certain definite temperature the thermal energy is insufficient to overcome the energy of interactions. This increases the viscosity of the polymer and decreases the intensity of thermal motion of its units.

METHODS OF DETERMINING T_g

The transition of a polymer from the rubber-like to glass state is accompanied by a gradual change in its physical properties and T_g can be determined by studying the variation of these properties with temperature. The most popular methods are:

(1) specific volume (dialatometric method)

(2) heat capacity

(3) modulus of elasticity

(4) deformation

The difference between a glass and the corresponding liquid or supercooled liquid may be demonstrated by the isobaric change in volume with temperature for an imaginary polymer as shown in Fig. 3.29 The polymer melt on cooling can form (a) a crystal or (b) a glass depending upon experimental conditions. If the substance crystallizes, *i.e.*, follows the path A, B, C, D, it is observed that as the liquid cools from any temperature T, to T_f, there is jump in the volume at T_f, the freezing point. Normally there is marked decrease in volume but in some cases like water, it increases. The coefficient of expansion of the liquid and of the crystal are constant in the vicinity of T_m. As the liquid is cooled slowly, its molecules arrange themselves into a definite pattern forming a crystal.

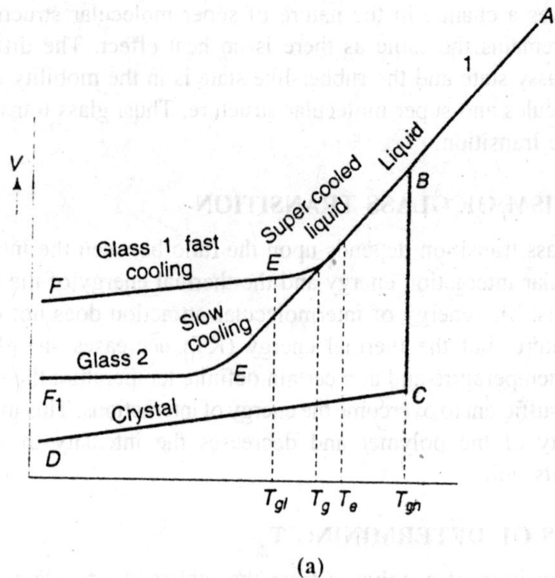

(a)

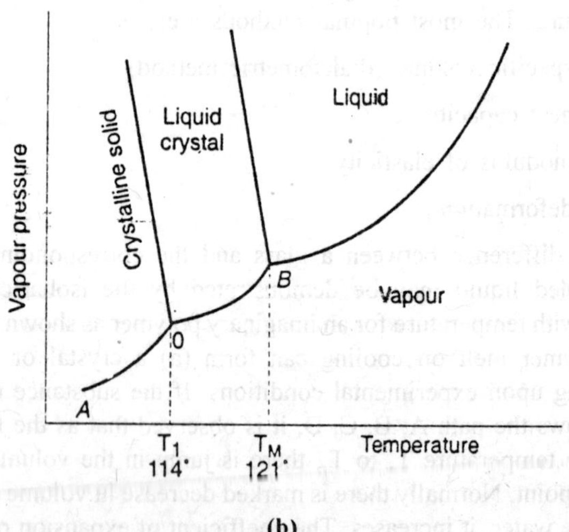

(b)

Fig. 3.29 : Volume-temperature curve for glass, supercooled liquid, liquid and crystalline state

In some of the liquids the forces of attraction (covalent or hydrogen bonds) between the molecules are so strong even above the freezing point, that they cannot freely move and do not provide a suitable crystal nucleus, the crystallization is prevented. When the liquid is rapidly cooled beyond its freezing point T_f the molecules do not have the time to organize themselves all over the liquid and hence, it is supercooled. The individual atoms or molecules are in an ordered array over a short range, but there is no long-range order. They are indeed in structure, very much like a snapshot of a liquid and are called supercooled liquid. Molecular movement in such liquids is minute and the attempt to describe the flow properties in terms of viscosity to the rigid glass is difficult. Below Tg*l* the dependence is again linear. Te is the start of the solidification and Tg*l* is the termination of the solidification. The temperature at the intersection of the extrapolated curves Tg is known as glass temperature. The molecules in the glassy state oscillate about their equilibrium position and their interchange of places either does not occur at all or only very rarely. The decrease in the thermal energy of the molecules with the fall in temperature decreases their ability to break down strong covalent bonds, when most of the motion is frozen at a temperature Tg*l*, bearing the liquid in a random and disordered state, a solid state results which is known as glassy state. One may say that the glass represent the metastable state of order of a liquid which is frozen in the range Te & Tg*l* and preserved at the lower temperature rather than a specific kind of material.

For the supercooled liquid in the range Tg and Te at a given rate of cooling (glass 1 A 2) the equilibrium distribution corresponding to each temperature is reached within the duration of the measurement, since the substance is in thermodynamical equilibrium with reference to arrangement and distribution of molecules. When a liquid on cooling changes to crystal, there is an evolution of latent heat associated with the abrupt change in the disorder, but there is no evolution of latent heat during the liquid to glass transformation. This suggests that the molecular arrangement in glass is similar to that in liquid. The glassy state is obtained by the inhibition of a kinetic process rather than a second order transition.

STRUCTURE

The structural units in glass are randomly linked and the regularity in the lattice has been found to extend to 5–10° A only. So it is evident that it has only a short-range order.

Silicates are known to exist in three crystalline and one amorphous form as shown below:

$$\text{Quartz} \xrightarrow{870^{\circ}} \text{Tridymite} \xrightarrow{1470^{\circ}} \text{Crystabollite} \xrightarrow{1710^{\circ}} \text{glass}$$

Silicon atoms are bound to four oxygen atoms forming a regular tetrahedron. The distance between the centres of the silica and oxygen ions is 1.6° A. In crystoballite (Fig. 3.30) the structure consist of a set of interlocking rings each with six atoms, three silicon and three oxygen held together by covalent bonds forming a giant molecule like that of carbon in diamond. All the three varieties soften at 1600° and melt to a liquid at 1700°. The silica crystal can melt only by breaking of the bonds. In the liquid a large number of Si-O bonds have broken to allow the parts to move about. However, the breaking is not irrevocable; Si-O bonds are continuously forming and breaking in the melt so that most of the time silicon is attached to four oxygen atoms and oxygen to silicon atoms. When the melt is cooled rapidly, the particular bond pattern present at that instant is frozen to a liquid of very high viscosity. Instead of a regular pattern of six membered rings, the tetrahedron is distorted resulting in jumbled rings of different shapes and sizes. The positional disorder existing within each distorted tetrahedron prevents the prediction and assignment of fixed position to the ions. The resulting glass is still a giant molecule but a disordered one. It is debarred from changing into a regular structure since that would involve the breaking and reforming of strong covalent bonds. The strong directional covalent bonds, present in the liquid, form wrong connections leading to glass. Although a short-range order is followed they form rings with the wrong number of elements. Moreover, the atoms in glass are at rest and there is a fairly constant distance between each atom and its -neighbour. The large interatomic distance do not have the same regularity as in the crystals. Therefore, the glasses have partially ordered material as is evident from two or more diffused haloes in the X-ray photographs. In view of very strong covalent bonds and very high viscosity of the liquids, the probability of the atoms to break their connections and to reform in the right direction is very low. Therefore, they form a disordered arrangement with strong interactions, resulting in glass. Since different energies will be required to breakdown the different meshes of structure, there is a continuous softening range instead of abrupt points.

Warren ruled out the possibility of the existence of molecules in the glass and found very good confirmation in the calculated and observedmaximas. It was very convenient to describe the change in viscosity on cooling to an association or polymerization process.

Zachiarasen deduced the following condition to be followed by an oxide glass of the composition $A_m O_n$: (1) No oxygen atom can be bound to more than two A atoms, (2) The coordination of oxygen around A must be low. (3) The oxygen polyhedra must not share edges or faces but only comets, and (4) At least three comers of the oxygen polyhedra must be shared.

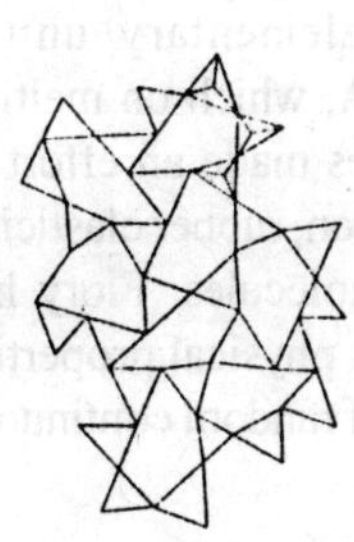

Fig. 3.30 : Quartz glass.

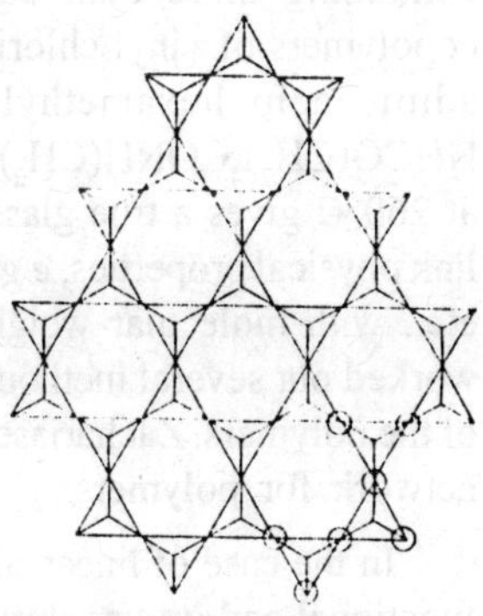

Fig. 3.31 Crystoballite.

There are two types of substances that follow the above conditions:

Those substances, that have directed covalent bonds, crystallize in an open structure, *e.g.*, SiO_2, B_2O_3, GeO_2, P_2O_2, As_2O_5, V_2O_5, Sb_2O_3, Sb_2O_5, As_2O_3, ClO_5, Ta_2C_5.

Substances whose liquids consist of long molecules which become entangled on cooling and require breaking of many covalent bonds for rearrangement, *e.g.*, plastic sulphur, organic glasses.

ORGANIC GLASSES

It is perhaps not always recognized that glass formation is a property shared by a very large class of compounds. These are polyphosphate glasses, those formed by certain elements in the polymeric form, *e.g.*, S, Se, those formed by hydrogen bonded materials like alcohols and organic polymers.

Lourenco obtained a highly polymerized viscous undistillable residue by polymerizing ethylene glycol in presence of ethylene dihalide, but

Staudinger was the first who suggested chain formulas for organic polymers. The polymeric glasses comprises a class of plastic materials of long chain organic molecules having an amorphous structure. The polymer chains consist of single repeating monomer units joined together by covalent bonds. Polanyi and Herzon suggested that no unit cell of crystal can be smaller than a single molecule. In 1929 Staudinger clarified the confusion regarding the structure of polymers.

However, certain plastics are also used as copolymers, *i.e.*, polymers containing more than one monomer unit, *e.g.*, styreneacrylonitrite copolymers of vinyl chloride, vinylacetate and nylon, synthesized from adipic acid hexamethylene diamine with an elementary unit—$NHCO(CH_2)_4CONH(CH_2)_6NH$—of a length of 17°A, which on melting at 260°C gives a true glass. Carothers and associates made an effort to link physical properties, *e.g.*, viscosity, double refraction, rubber elasticity, etc., with molecular weight and shapes of these molecules. Flory has worked out several methods to determine the various physical properties of the polymers. Zachariasen propagated the theory of random continuous network for polymers.

In the case of linear polymers each chain molecule retains only two functional end groups during the process of polymerization irrespective of its size. The main chemical type of polymer glass may be represented by polystyrene, polymethyl methacrylate, polyvinylchloride, and polycarbonates, etc. Thus, polystyrene normally contains a small quantity of lubricant which reduces the glass transformation temperature by 10°C and improves its moulding properties.

In nonlinear polymers the number of functional groups increase for each molecule with its size or us molecular weight, *e.g.*, formation of various sulphur, where S_8 ring opens and produces a chain between two molecules by the pairing of electrons.

Polymethyl methacrylate is a glass below 90°C. The mixture of polymer and monomer, or of polymer, monomer and the diluent, transform to a glass. During the process of polymerization, the percentage of polymer reaches a certain well defined high value. Even the monomer molecules are frozen up in their position in the glassy state.

The following organic substances have been found to form glass. Ethyl alcohol, propyl alcohol, glycerol, glucose, water, rubber and silicones where –OH is the responsible group. Toluene where Van der Waals forces and covalent bonds are responsible for glass formation.

Table 3.11 : Monomers forming glass and their T_g

Monomer	T_g *in °C*
Vinylchloride	75
Vinylacetate	30
Styrene	100
Methylacrylate	0
Methyl methacrylate	90
Methyl acrylonitrile	120
Vinyl isobutyl ether	–20
Isobutylene	–70
Butadiene	–88
4-Methyl pentene	2
3-Methyl-1 butene	23
Ethylene terephthalate	37
Acrylonitriles	73

Recently, the organosilicone compounds, because of their greater stability, have become important. The Si-CH_3 or Si-C_2H_5 bonds have greater oxidation resistance than the corresponding C-C bonds. Depending upon the molecular weight, chain length, ring structure and the degree of three-dimensional polymerization, the methyl or ethylsilicones can have very different properties ranging from fluid to viscous oil, resins and elastomers. The products obtained depend on the methods of preparation.

Warren, Bagchi, Alexander and Miller, have discussed and given upto date account of the application of X-ray to glassy materials. Debye, Station and Heikens have used the X-ray scattering method for determining the void spaces in polymers and glasses. Interesting relationships have been observed between the heterogeneity of the glass and mechanical properties of fibres. Several workers have also applied NMR, ultrasonic and diffusion techniques for the above purposes.

The constitution of an organic polymer is guided by the following parameters:

1. The configuration describes the atomic structure of the molecule according to the chemical formula. The configuration is the same in solution, in the glassy state or the crystalline state.

2. The conformation is a dynamic parameter, *e.g.*, the rotation of the segments about a single bond. They may have different number of elementary units per turn of the helix but in every case the number of possible conformation will be greater in liquid or in the glassy state than in the crystal.
3. The molecular packing or chain packing refers to the lateral organization of the chains in the crystal. Since most of the polymers even as crystals are only poorly defined, this parameter covers different types of disorders and lattice imperfections.

Du Pont have reported new polymers called ionomers by crosslinking chains of polyethylene through the ionic forces acting between (–COO) groups and Mg^{+2} or Zn^{+2} as cations. Rees and Vaughan have been successful to introduce strong ionic forces between the chains through their carboxylic groups. The ionomers promise a rare insight into the effect of gradual transition of Van der Waals forces and hydrogen bonding into the three-dimensional ionic network. The ionomer resins are solids with a high vibrational entropy, introduced by the electrical fields, screening demands and nucleating effects of cations which interfere with the establishment of long-range order. All these facts result in the following properties:

1. The stiffness increases with the acid level and the concentration of cations.
2. The ionomer resins exhibit unusual stability in high voltage application.
3. The ionomer resins have unusual elasticity and toughness.
4. The ionomers, because of association in liquid state resulting in high viscosity, behave like high molecular weight thermoplastics than like cross-linked polymers.
5. Compositions containing dibasic acids in conjunction with di- or trivalent ions resemble cross-linked olefins in the behaviour of the melt.
6. The temperature coefficient of viscosity is very sensitive to the strength of bonding forces.

Covalent C-C bond in cross-linked polymers present flow at temperatures at which the polymer is thermally stable. Ionic bonds and resulting association in ionomers makes the viscosity of the ionomer very sensitive to temperature giving activation energy of the order of

20 kcal/mole.

Hydrogen bonding through carboxylic acid molecules makes the viscosity less sensitive to temperature where the activation energy is 14 kcal/mole. Van der Waals forces in the polymers show the lowest temperature coefficient, *i.e.*, 7-19 kcal/mole.

The intermolecular binding forces determine its solubility in organic solvents and permeability for oils, Ionomers have extremely low permeability for oil and are insoluble in common solvents. Special mixtures, *e.g.*, N, N-dimethylacetamide, $CH_3CON(CH_3)$; added to decalin can dissolve ionomer.

Rose and Block have synthesized some inorganic polymers by refluxing dibutylphosphoric acid with zinc acetate.

The above polymer remains flexible down to –80°C. It is soluble in chloroform and a polymer of molecular weight 10,000, melts at 150°C and forms a glass on cooling.

GLASS TRANSITION TEMPERATURE AND MOLECULAR MASS

Glass transition increases rapidly with the increasing mass and finally reaches a constant value (Fig. 3.32). The molecular mass at which T_g becomes constant depends on its chain flexibility. For isobutylene T_g becomes constant at M equal to 1000 while polystyrene M should be more than 15,000.

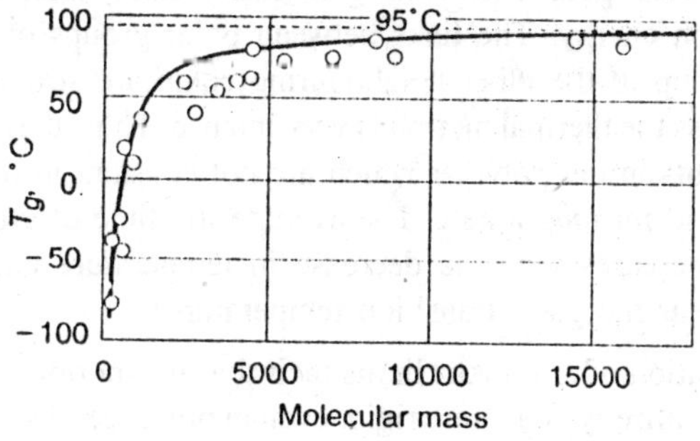

Fig. 3.32

Low molecular weight polymers exist only in two states, *i.e.*, the glass and the liquid and the mass transition temperature coincide with the flow temperature. But, with increasing molecular weight at a certain molecular mass the transition temperature splits into T_g and T_f, thus, introducing a rubber-like state. With further increase in mass, T_g becomes constant but T_f continues to increase. Thus, $T_f - T_g$ gives the temperature range of rubber-like state.

It has been experienced that low molecular mass polymers exist as glassy or liquid where their T_g and flow temperature coincide. With higher molecular mass $T_g - T_g$ is great showing the increase in rubber range. The molecular mass at which the transition temperature splits into T_g and T_f depends on the chain flexibility. The more rigid the chain, higher will be the molecular mass at which the T_g splits. Thus, in polyisobutylene rubber-like strain appears at molecular mass 1000 while in polystyrene, it appears at M 40000.

Polymers with less flexible chains have very high T_g and they do not display rubber-like structure even at higher temperatures. In such case, we talk of softening with a range of temperature. T_f of amorphous polymers increases with increasing molecular mass.

These fact go a long way in the synthesis of polymer with desired properties.

In the case of polydisperse polymers, polymer fractions of different molecular mass pass into visco-fluid state at different temperatures.

GLASS TRANSITION TEMPERATURE AND CHEMICAL CONSTITUTION OF POLYMERS

Polymers with polar molecules give easily glassy state because of high interaction energy. The links between polar groups of one chain with polar group of the other neighbouring chain are stronger and do not break unless the thermal motion is very intense. Thus, they form local cross-link points in the polymer which are not constant in time and go on breaking and forming a new. The average life time of groups in the linked state increases with the decrease in temperature and becomes appreciable near the glass transition temperature.

The formation of crosslinks limits their thermal motion. This results in making the entire system more rigid, even in presence of small number of crosslinks, forming a structure of fixed random arrangement of polymer molecules. The polymer acquires the properties of a solid below T_g. It has also been observed that in glassy polymers most of the hydroxyl groups takes part in hydrogen bonding.

Branched-chain and linear aliphatic resins provide low crosslinks and higher chain flexibility. One group on a given molecule with more than one reactive site on the adjacent molecule tends towards cyclization process rather than crosslinking. The effect of the types of linkages and T in the case of epoxy resin.

FORCED ELASTICITY

Low molecular glasses like resin and silicate glasses do not respond to large stresses. Macromolecular glasses often becomes brittle at temperatures much below T_g. The ability of glassy polymers to deform considerably without breaking is responsible for their wide use in engineering. The large deformation of glassy polymers was at one time attributed to flow process called cold flow, however, mutual displacement of macromolecules is not probable.

Methyl methacry late under a strain at temperature below T_g was found to recover its initial shape and size after heating above T_g. This recoverable nature of large deformation of high molecular glasses suggests that the regularities of rubber-like state apply to the glass of state as well, showing that the glass transition is not a phase transition and a polymer possesses the same structure at temperatures above and below T_g. The relaxation time of polymer in glassy state is very great.

Equation 1 gives a relation between relaxation time and temperature

$$\tau = \tau_o \, C^{\Delta u - \alpha s/kT}$$

where τ is the relaxation time, ΔU is the activation energy, a is the stress and a is a constant.

At small stresses, stress value does not affect relaxation time but at larger stresses τ becomes smaller and highly elastic strain appears. It can also be seen that after the removal of deforming stress, elastic strains do not disappear at temperatures below T_g.

STRESS STRAIN DEPENDENCE

Glassy polymers like polystyrene, poly (methyl methacrylate), poly (vinyl chloride), etc. on deformation begins to extend non-uniformly when a particular stress is reached. One part of specimen becomes thinner than the rest of the mass and a neck is formed. Glassy polymers like cellulose acetate, cellulose nitrate do not form neck upon forced elastic deformation. The stress at which the total rate of deformation becomes equal to force elastic deformation is called ultimate forced elasticity.

The ultimate forced elasticity has been found to rise with increasing straining rate indicating its relaxation nature.

BRITTLE TEMPERATURE

The forced elasticity not only depends on temperature, but also on the rate of stress applied. The temperature where the stress dr becomes zero, rubber-like deformation develops. At a certain low temperature, the stress required to regroup the chain sections correspond to so short a time that σ_f reaches the brittle strength values σ_{br}. The temperature below which the polymer breaks under the stress σ_{br} is known as brittle temperature.

The brittle temperature can be obtained by plotting curves between σ_f vs T and σ_{br} vs T. The intersection of these two curves will give brittle temperature (Fig. 3.33).

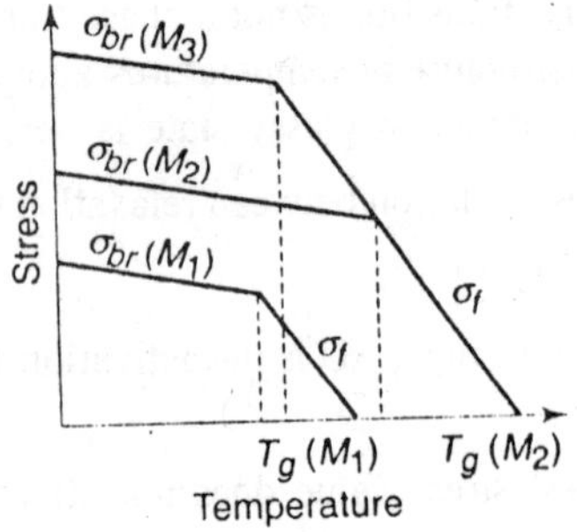

Fig. 3.33 : Temperature dependence of brittle strength and ultimate forced elasticity

Macramolecular glass even on being subjected to the same rate of deformation have different T_g and T_{br}. The difference $T_g - T_{br}$ determines the temperature range of forced elasticity.

Below T_{br} large stresses will cause brittle failure of polymers. Glass polymers give the best performance between T_g and T_{br}. This makes a large temperature range of forced elasticity, a very valuable property in a polymer. The range of forced elasticity of some of the glass polymers.

Thus, Table 3.12 shows that the forced elasticity ranges of different polymers vary widely. The forced elasticity range depends on T_{br} which depends on T_{br} and on the nature of change of σ_f with temperature.

The dependence of d is determined by intermolecular attraction energy, packing in macromolecules and the molecular mass of the polymer.

Table 3.12 : Temperature range of forced elasticity

Polymer	T_g °C	T_{br} °C	$T_g - T_{br}$
Polystyrene	100	90	10
Poly (methyl methacrylate)	100	10	90
Poly (methyl 100% extended)	110	–50	160
Poly (vinyl chloride)	81	–90	171
Natural rubber	–62	–80	18
Butadiene rubber	–39	–112	73
Styrene based rubber	–49	–135	86
Nitrile rubber	–19	–110	91

EFFECT OF INTERMOLECULAR ATTRACTION ENERGY

In polar glass, macromolecules form strong bonds by crosslinking between the polar groups of the chains which increases its brittle strength. These bonds are also labile and large stresses cause a regrouping of chain sections to load the entire molecular network more uniformly. Thus, the forced elastic deformation occurs at lower temperatures.

EFFECT OF PACKING IN MACROMOLECULES

The loose packing of some of the macromolecules favours the forced elasticity to appear on application of large stresses. On cooling, σ_f does not change appreciably and the brittle temperature is low, *e.g.*, the loose packing of macromolecules of cellulose nitrate and cellulose acetate causes a wide range of forced elasticity.

EFFECT OF MOLECULAR MASS

The molecular mass affects the brittle strength and T_g of the polymer. T increases with molecular mass of a polymer up to n = 200 (n is degree of polymerization) and the brittle strength up to n = 600. With increasing n both T_g and σ_{br} increases up to n = 200. With further increase of n,

T_g remains constant, but σ_{br} continues to increase and T_{br} decreases (Fig. 3.33).

Above the flow temperature T_f is the viscofluid region, between T_f and T_g is the rubber-like state region, between T_g and T_f is the forced elasticity region and below T_{br} the polymer is in the brittle state.

GLASS TRANSITION BY CHROMATOGRAPHY

Partition of solute molecules between gas and stationary phases, expressed in gas chromatography as a retention volume, is a characteristic of the system investigated. A plot of the logarithm of the retention volume versus reciprocal of absolute temperature should be a straight line whose slope is directly related to the corresponding enthalpy, heat of solution in gas-liquid chromatography (GLC) and heat of adsorption in gas—solid chromatography (GSC). Polymer stationary phases deviate from such linear behavior at their glass transition temperature (T_g), leading to a z-shaped retention- diagram as depicted in Fig. 3.34. It has been shown that a change of retention mechanism occurs at T_g. While below T_g (region I) retention proceeds exclusively by surface adsorption, increasing molecular mobility at and above T_g (region II) is accompanied by an increase in retention volume due to bulk sorption. At even higher temperatures (region III), equilibrium conditions are restored, and bulk retention volumes decrease with increasing temperature. Because of the high sensitivity of GC detectors, measurements can be made with very low concentrations of solutes at a level where they do not alter the structure of the polymer under study.

Due to the occurrence of two retention mechanisms above T_g the shape of the retention diagram is dependent on experimental conditions. When both surface adsorption and bulk sorption are operative in a column, equilibrium retention volumes are satisfactorily represented for low molecular weight liquids and polymers by the relation:

$$V = K_s W_L + K_a A_L$$

where V is the total retention volume, K_s and K_a the partition coefficients for bulk sorption and surface adsorption respectively, W_L and A_L the mass and surface area of the stationary phase. As the coating thickness (W_L/A_L) is decreased, reversal from the linear behaviour becomes less pronounced, and at low enough coverage a linear retention diagram through T_g is obtained, corresponding exclusively to surface adsorption. The temperature of first deviation from linearity T_1, representing the onset of bulk sorption, was found in excellent agreement with the glass transition temperature of the stationary phase, independent of experimental conditions and thickness of the stationary phase. This is consistent with the concept of the glass transition as an iso-free volume state. The penetration of the problem molecule into the bulk of the polymer provides a sensitive test for the rapid development of excess free volume which is known to occur at or just above the glass transition temperature.

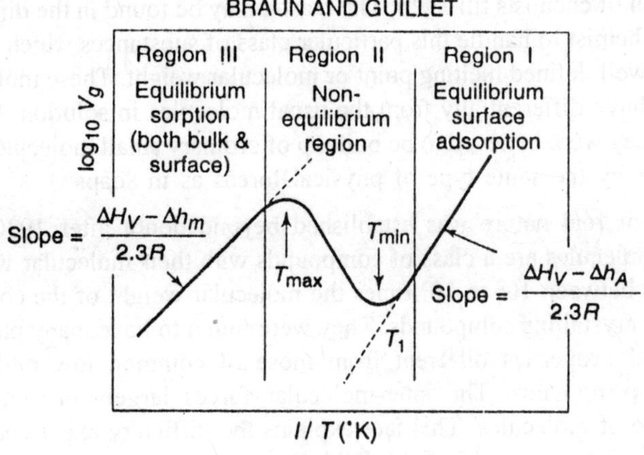

Fig. 3.34

Retention volumes were measured at intervals of approximately 3°C with n-pentane on a sample of isotactic polypropylene coated onto an inert support. In view of the temperature range of interest, –40 to + 50°C, a low boiling probe was necessary to produce suitable retention times. As was the case with amorphous and rubbery polymers, a large and characteristic change in retention behavior was recorded for this highly crystalline (62%) polymer in the glass transition region. Results are in good agreement with published transition temperatures of-polypropylenes of similar density, as can be seen.

PLASTIC STATE

There is no product so widely used in every sphere of life than polymeric materials, designated as resins, resinous materials, plastics and elastomers. They are generally high molecular weight compounds comprising of macromolecules. The earliest reference to the term resin implies natural materials produced by vegetable or animal life, *e.g.*, shellac. A reference to *'laksa'* has been found in *Atharva Veda* and its use as a moulding material is given in *Mahabharat*. As a matter of fact the word shellac, is derived from silact.

High polymers or macromolecules comprise the most widely distributed kind of naturally occurring organic materials in the form of starch, plant fibrous material (cellulose), proteins and rubber. It appears Strange that though nature was so successfully synthesizing these materials so important for the life of any living creature, it could not attract the attention of chemists till 1920. The reason may be found in the difficulty of the chemist to handle this particular class of substances which do not have a well defined melting point or molecular weight. These molecules also behave differentially from the usual molecules in solution. Before 1930, they were believed to be built up of ordinary small molecules held together by the same type of physical forces as in soaps.

Their real nature was established beyond doubt after 1930. The macromolecules are a class of compounds with their molecular weights ranging between 10^2 to 10^5 times the molecular weight of the common organic crystalline compounds. They were found to have many physico-chemical properties different from those of common low molecular weight compounds. The inter-molecular forces largely influence the structure of molecules. This fact explains the difficulty experienced by the early workers in this field. With the establishment of the fact that polymers are constituted of macromolecules, the chemistry of macromolecules advanced at rapid stride and the tremendous developmant in this area was responsible for the production of synthetic films, fibres, elastomers, plastics and inorganic polymers. The clear understanding of these substances has helped in the development of the new branch of biopolymer substances that play a very vital role in living cells.

The electro-chemical studies of proteins by Tiselius proved that the amount of free charge, which the colloidal particles possess, plays an important role in influencing their stability. Tiselius received Nobel Prize in 1948. Recently the brilliant synthesis by Natta and co-workers of 1, 3-dienes to exclusive chain configurations have opened the way to a new approach to synthesis. He has synthesized structures identical with natural rubber and balata rubber.

The advancement in the fields of physics and chemistry and with the development of X-ray diffraction diagram and refined methods of determining the size and the molecular weight of macromolecules, it can be safely concluded that the plastic state is also a state of matter.

PLASTICS

Substance which flow under heat or pressure or both, are known as plastics. They do not include clay, glass and metal. They are mostly organic in origin and can be softened by heating and moulded in this condition. Resin is their essential constituent.

They do not have, like glass, a defined melting point, molecular weight, etc. These odd class of compounds is characterized by the presence of units, molecules ranging 10^3 to 10^5 time the molecular weight of common crystalline organic compounds. They possess a marked range of deformality (100 to 200%) particularly at high temperature which is partly reversible and partly permanent. They possess moduli of elasticity in the range of 10 to 100 kg/cm^2.

CLASSIFICATION ACCORDING TO MECHANICAL PROPERTIES

Polymers according to their mechanical properties have been classified as elastomers and plastomers. Elastomers are rubbery elastic materials. Plastomers or plastics are macromolecular materials which flow on deformation. The deformation is permanent in plastics but in elastomers the body returns to its initial state instantaneously upon the removal of stress. Normally, in elastomers the chains are very flexible at the room temperature but they are stiff in the case of plastics. This depends on several factors namely, the degree of polymerization, cohesive forces acting between the chains, flexibility of the chains, their branching and crosslinking and the degree of crystallinity.

RESINS

There is no explicit definition for resins. A resin is generally understood to be a solid or a semisolid (very viscous) at the room temperature. It is a natural or synthetic organic substance of relatively high molecular weight which exhibits no sharp melting point and usually is predominantly amorphous in structure. It is normally insoluble in water but soluble in some organic solvents. This is made up of organic molecules, which are held together by some kind of physical forces.

CRYSTALLITE

When two or more long chain molecules come close enough to laterally look in any way, the nucleus of crystallite is born and the crystallite begins to grow. It consists of a cluster of such associated chain molecules which form the basis of structural unit. The longest dimension of these micro-units in a polycrystalline material vary in size from 50-500 A. These long chain molecules are so tightly packed that even X-ray beams are diffracted from them, which is also the basis of finding the size of the crystallite.

MICELLES

They are aggregation of the crystallites of colloidal dimensions that exist in the solid state or in solution as spherical or rod-like shapes. They represent a reasonably reproducible dimension as a result of uniform chemical or mechanical treatment such as soaps or colloidal electrolytes. It has been observed in plastics that these long chains wander successively through disordered, random regions, through bundles of organized regions (miscelles), again amorphous, again ordered and so on. This fringed micelles model is very useful in explaining the oriented structure of drawn fibres.

MONOMER

It may be a molecule or a distinct segment of molecular chain consisting of either a molecular unit or group of molecular units, but it contains the same kind and number of atoms as the real or hypothetical unit. It may be either of organic or inorganic origin, *e.g.*, formaldehyde, styrene, sodium phosphate, etc.

POLYMERIZATION

The process of joining together small molecules or monomers by covalent bonds to form large molecules, with or without formation of other products is known as polymerization, *e.g.*, paraformaldehyde from formaldehyde. The actual mechanism by which some naturally occurring substances like rubber are formed is not completely understood. The ultimate starting materials like petroleum, natural gas, coal tar and cellulose are few in number.

We may classify the various processes involved in polymerization as follows:

1. By the number of bonds each monomer can form in the reaction.
2. By the kinetic scheme governing the polymerization reaction.
3. By the chemical reaction used to form new bonds.
4. By the number of monomer used in polymerization.

5. By the physical arrangement, *e.g.*, conformation.

The most important classification by Crothers & Flory is as follows:

1. Stepwise or condensation polymerization.
2. Chain or addition polymerization.

In stepwise polymerization each group is as reactive as the monomer. While in chain polymerization each group becomes dead and can no longer react with the monomer.

Degree of Polymerization

The degree of polymerization (D.P.) is the average number of repeating units in a linear polymer. If such a polymer consist of regular repeating monomers or units, the average number of monomers per polymer is called the degree of polymerization. It is obtained by dividing the average molecular weight of the polymer by the molecular weight of the monomer. The average molecular weight of poly vinyl chloride is 63,000. Therefore, D.P. is 1000.

OLIGOMERS

The chain like molecules with a few (5–10) structural units are called oligomers. 'Oligos' means few and 'meros' means parts. These are low molecular weight products. Many organic compounds are of fairly high molecular weight, yet they are not considered as polymers. It is the complexity of their molecules and not the multiple repetition of the monomer which accounts for their molecular weight and size.

POLYMERS (MACROMOLECULES)

Staudinger [3] suggested the term macromolecules for polymers (Poly means many). They are giant molecules obtained by the chemical reaction between the monomers. The monomers are held together by covalent bonds while the separate molecules are attracted to each other by Van der Waals forces. They may be partially crystalline and partly disordered or completely disordered as in glassy or brittle state. Very often they differ from the original structural unit in molecular weight

$$n(CH_2O) \rightarrow (CH_2O)_n$$

STRUCTURAL UNIT OR RADICAL

It is very important to have a clear understanding of structural units. The structural units are not monomers by themselves but are free active groups or radicals generated in the course of reaction. Table 3.13 clearly indicate them.

Table 3.13

Polymer	*Monomer*	*Structural Unit*
1. Polyethylene	$CH_2 = CH_2$	$-CH_2\ CH_2-$
2. Polyvinylchloride	$CH_2 = CHCl$	$-GH_2\ CHCl-$ |
3. Polysiobutylene	CH_2 | CH_3-C-CH_3	CH_2 | CH_3-C-CH_3 | H
4. Polystyrene	H | C_6H_5-C- | CH_2	H | C_6H_6-C | CH_2 |
5. Poly sebasic anhydride	$HOCO\ (CH_2)_8\ COOH$	$-OCO\ (CH_2)_8\ -CO-$
6. Polyhexa-methylene-adipamide	$H_2N(CH_2)_8\ NH_2$ + $HOCO\ (CH_2)_4\ COOH$	$-NH-(CH_2)_6-NH-CO-$ $(CH_2)_4-CO-$
7. Phosphonitrilic chloride	PCl_5 + NH_4Cl	Cl | $-P = N-$ | Cl
8. Silicon disulphide	Al_2S_3 + SiO_2	S / \\ > Si < \\ / S
9. Sulphur trioxide	SO_3	‖ O ‖ O ‖ –S–O– ‖ O

In the first four examples the structural units and the monomers from which they are derived possess identical atoms occupying similar relative positions, only a rearrangement of electrons takes place during polymerization.

In the two examples 5 and 6 water molecules were eliminated during polymerization process. In the last case it is convenient to define as structural unit, the pair composed from both possible structural units.

FUNCTIONALITY

The average number of reactive groups or the number of bonds produced per monomer molecule will decide the nature of polymerization. Carothers defined this number of bonds as the functionality of the monomer.

Monofunctional monomers give only low molecular weight products, *e.g.*, acrylic acid gives ester or amide. Bifunctional monomers give linear polymers, *e.g.*, butadiene. Trifunctional monomers give non- linear polymer. Glycerine is a trifunctional monomer, which on condensation with pthalic acid, a bifunctional monomer yields a non-linear polymer. If the condensation proceeds far enough, a network structure is easily obtained.

The linear or non-linear polymers are based on the absence or presence of polyfunctional units. In practice linear polymers do possess some limited branching because of unspecified side reactions intervening during polymerization, *e.g.*,

$$CH_2 = CH\text{–}CH = CH_2 \rightarrow (CH_2\text{—}CH = CH\text{–}CH_2)n$$

$$CH_2 = CH\text{–}CH = CH_2 \rightarrow \begin{array}{c} H \\ | \\ \text{–}CH_2\text{–}C\text{–} \\ | \\ \text{–}CH_2\text{–}C\text{–} \\ | \\ H \end{array}$$

It is also known that many non-linear polymers consist mainly of linear sequence units which are interpreted only at rather wide intervals by units of high functionality. This relatively small fraction of functional units may bring about profound changes in certain physical properties of the whole polymer.

HIGH POLYMERS

The term high polymer is indefinite in the sense that the minimum size or the length of the molecule is unspecified. It is intended to apply to very large molecules of such a size that the properties associated with long chain molecules become evident. Their chemical and physical structure depends on the arrangement in sequence of many monomers connected by primary chemical bonds to form long chains. They are the multiples of monomers which may not be of the same size or same composition or chemical structure. Staudinger coined the word macromolecules, while in Great Britain and United States the term high polymer is used, but both have the same meaning. Cellulose, protein, starch, rubber and shellac are some of the natural high polymers but the first two are most widely distributed in nature. Cellulose form the chief component of the cell walls of all plants and proteins which are the essential constituent of living cells.

Most of these high polymers have molecular weight in the range of a few thousands or millions.

Primary Structures

The primary structures are produced by the polymerization of chemically identical monomers into long chains without subsequently changing its chemical constitution or configuration, but they do affect their solubility and chemical behaviour by influencing the degree of flexibility or mobility of the chain. These are c-c chain polymers.

Secondary Structures

The secondary structure relates to the type of aggregation state of the polymer chain of a single macromolecule. It can exist in different shapes, *e.g.*, extended chain, random coil, periodically regular arrangement of chain segments.

In solution the average coil density is proportional to the viscosity and the molecular weight, *e.g.*, proteins, D.N.A.

Tertiary Structure

It relates to the arrangement of the macromolecule to form more complex aggregates, *e.g.*, cell structure of random coils, spaghetti structure, fringed micelle, spiraling chain, etc.

It can extend uniformly throughout the whole polymeric material and may lead to quarternary structures like haemoglobin.

Derived High Polymer

When a primary high polymer or a natural high polymer is altered chemically so as to produce a derivative, it is called a derived high polymer, *e.g.*, cellulose nitrate, sodium salt of nylon, etc.

TYPES OF HIGH POLYMER

The justification for such a division lies in the net difference between the processes by which polymers may be synthesized, viz. condensation and addition reactions. According to the first mechanism polymers are formed by stepwise intermolecular condensation of monomer functional groups with the elimination of a bye product while the last two takes place by chain mechanism involving active centres.

Condensation (Stepwise) High Polymers

It is a special type of polymerization commonly called C-polymerization. Flory stresses more on the mechanism by which the polymer is formed. They are formed by the inter-molecular condensation of the reactive groups. In this a union between pair of functional groups of the monomers takes place involving a chemical change with the formation of an inter-unit functional group not present in the monomers and also releases or splits off simple molecules like water or ammonia coincident with the formation of macromolecule. The monomers disappear at an early stage says 99% at D.P. Any two species of molecules can react to give products with molecular weight rising steadily throughout the reaction. A polymer can be prepared in two ways.

The class of high polymers which on chemical degradation yield monomeric products differing in chemical composition from the structural units are also called condensation high polymers.

Space Network High Polymers

When there are two or more reactive functional groups in a monomer or monomers, the chemical reaction proceeds first in two dimensions and then in three dimensions. This is known as a space network high polymer, *e.g.*, glycerol is a trifunctional monomer, which on condensation with bifunctional phthalic acid yields a non-linear polymer, which on further condensation proceeds to give a network structure.

Addition (Linear) Polymer

Addition polymerization usually takes place by a chain mechanism involving active centres of one sort or another and results in polymers in which the chemical formula of the structural unit or units is identical with the starting monomer. Each polymer, formed in comparatively short time, becomes dead and remains unchanged by the reaction of the remaining monomer. The most important class of addition polymers comprise those derived from unsaturated monomers. The second electron pair is held about 70% as firmly as the first which leads to high polarizability and chemical reactivity in unsaturated compounds. If an atom or group of atoms is substituted for a hydrogen atom adjacent to a double bond, the electrostatic force exerted by the electrons of the substitutent on those of the double bond affects the reactivity of the molecule. The macromolecule in solution have a negative charge, the small positive hydrogen ion come in close proximity of chains because of electrostatic attraction due to high charge density associated with the chain.

Many compounds having a particularly reactive double or triple bond may, when pure, be easily polymerized by means of an addition process. Most polymerization of the unsaturated compounds proceed by a free radical mechanism. Usually referred to as vinyl type of polymer, CH_2–CHX, where X may be a group such as phenyl, –OH, –Cl, COOH, For any ion which is able to group which on dissolving the polymer in a polar liquid dissociates leaving fixed charges along with the chains, *e.g.*, polymethacrylic acid. Addition polymerization in general exhibit some distinctive features.

1. Ready susceptibility to catalysts, photoactivation and inhibition.
2. The active centre of the kinetic chain is retained by a single polymer molecule throughout the course of its growth, *i.e.*, the individual polymers grow to maturity while most of the monomers are still unreacted.
3. Very high molecular weight polymers may be easily formed up to several millions with reaction time as low as a fraction of a second.
4. The overall polymerization involves at least two other processes, *i.e.*, chain initiation and chain termination.

The process of initiation may be promoted by the attack of a radical or an ion on the monomer with the formation of a new radical, carbanion or carbonium ion.

Let A be the repeating unit, then the linear polymer will be represented as

$$A–A–A–A–A \quad ...(8)$$

$$CH_2 = CH_2 \rightarrow –CH_2–CH_2– \rightarrow –CH_2–CH_2–CH_2–CH_2 \quad ...(9)$$

Atoms other than carbon also give similar structures such as sulphur, polyphosphates and phosphonitrile chloride, etc.

```
                         O       O       O
                         ‖       ‖       ‖
—S—S—S—S—    —O—P—O—P—O—P—
                         |       |       |
                         H      OH      OH

          Cl     Cl     Cl
          |      |      |
        —P—N—P—N—P = N—
          |      |      |
          Cl     Cl     Cl
```

Copolymer

High polymeric substances, containing two or more structural units combined more or less in a random sequence, are known as copolymers. The alternating sequence composed of two birfrictional units give a series of possibilities as shown below:

The term copolymerization indicates that the mixed monomers continue in the same polymeric chain. In case the monomers polymerize individually to form molecularly distinct species, each containing a different unit, then a mixture of homopolymers results rather than a copolymer.

Branched High Polymer

The long chain molecule in this type of polymer is not uniformly straight but has branches extending its trunk. The chain molecules remain unattached to other similar surrounding molecules. The synthesis of chain like macromolecules always occurs through the reaction of bifunctional components with each other. The bifunctional character of the monomer arises only with the addition of the initiator or ions, *e.g.*, vinyl compounds which in presence of a radical add to each other. Wherever chain transfer is involved in the formation of high polymer, each transfer reaction adds a branch point to the molecule. Since the probability of branching is proportional to the weight average molecular weight, so with the increase in molecular weight with the formation of branch, it tends to become more highly branched and larger.

In many systems a polymer can be dead to the usual addition of monomer, but alive to act as a transfer agent by different mechanism.

Cross-Linked High Polymers

The presence of compounds with more than two functional groups lead to the formation of branched and crosslinked structures, *e.g.*, polyesters, polystyrene in presence of divinyl benzene. They have long chain molecules bound together by cross bridges as crystalline bridge in wood.

The classification of a compound as branched or cross linked depends on the average degree of polymerization (D.P.) Generally as long as the compound is soluble, it is known as branched polymer. However, when it becomes insoluble in all the solvents, it is considered as a crosslinked polymer. In crosslinked polymer all or most of the chains of high polymer are rigidly tied to each other by covalent bonds.

The branched polymer can be converted to a three-dimensional crosslinked polymer, first through the increase in the degree of polymerization of the linear chains and second through the increase in the number of branching and crosslinking points.

Similarly, a crosslinked polymer can be converted into a branched polymer by exposing it to strong mechanical degradation.

Crosslinking produces both desirable and undesirable properties. It is a useful reaction in making a rubbery material which is stable at high temperatures. Crosslinked polymers can be used for high temperature electrical insulation. The rubbery system, on further crosslinking with the passage of time becomes brittle. It shrinks and its compatibility decreases.

IONIC POLYMERIZATION

Ionic polymerization can be initiated through acidic or basic compounds, *e.g.*, $AlCl_3$, BF_3, $TiCl_4$, etc. All of them are strong electron acceptors which are supplied by the double bonds of monomers. The initiation reaction of ionic polymerization needs very small activation energy hence, it can occur at temperatures below 50°C with explosive violence, *e.g.*, polymerization of styrene at –70°C in tetrahydrofuran. The chain termination in ionic polymerization takes place only through impurities.

BLOCK COPOLYMER

This is a special type of copolymer. The monomers consisting of segments or blocks of similar monomers are held together in the chain. This contains polymers with long sequence of two monomers which can have an arrangement with the sequence following one another along the main polymeric chain as

—AABB—BBAA—AABB—

In case of copolymerization of styrene and methyl methacrylate with ethylamine as the transferring agent, we obtain a block copolymer.

$$\begin{array}{c} N(C_2H_5)_2 \\ | \\ —AA—A—C—B—BB \\ | \\ CH_3 \end{array}$$

where, A is methyl methacrylate and B is styrene.

If a polymer of one kind is dissolved in the monomer of another and the mixture is subjected to polymerization, chain transfer between growing chains of the monomer and the molecules of the polymer form branched copolymers in which long chains of one type of unit are joined to a backbone chain consisting of the other unit.

POLYMER CRYSTALS

Polymers are products of high molecular mass and their decomposition temperatures are far below their boiling points. Hence, they exist only in liquid or solid state. Although most of the polymers are amorphous, the crystalline polymers known as isotactic polymer have high melting point and produce fibers of high tensile strength and excellent mechanical properties. A polymer has two types of structural elements viz., monomer units and chains. The long range order is a prerequisite for the crystallization of polymer, not only in the arrangement of units but of chains as well, in three dimensions. This can happen in three ways:

1. Long-range order of arrangement of both chains and monomeric units in three dimensions.
2. Long-range order of arrangement of chains, but the monomeric units are arranged disorderly.

3. Long-range order of the arrangement of units but short-range order of chain arrangement.

The high degree of ordering can be achieved in polymers in two ways, *i.e.*, by crystallization or by chain orientation without orientation of monomeric units. These correspond to crystalline and amorphous (glassy) state. It is possible to have a perfect short-range order in chains. But the long-range order in arrangement of chains and monomeric units is not very perfect. Therefore, the examples of well ordered liquids and defective crystals are well known.

CRYSTALLIZATION OF POLYMERS

Many polymers are incapable of crystallization under any condition. Their crystallizability is determined by their chemical constitution, *i.e.*, their packing, energy of intermolecular reaction, chain regularity, flexibility, etc.

Regularity of Polymer Chains

A long-range order must exist in the chain in three dimensions. Sometimes irregularities in the chain may not obstruct crystallization but random and atactic copolymers cannot be crystallized under any circumstances. Natta by using certain catalysts succeeded in synthesizing several higher crystalline vinyl polymers, *e.g.*, polypropylene, polybutylene and polystyrene.

A mixture of the polymers of various degree of orientation was formed and crystalline fraction was seperated from amorphous fraction by means of solvents, as crystalline polymers are less soluble than amorphous polymers.

Flexibility of Polymer Chain

Polymers with sufficiently flexible chain can crystallize. Rigid chain polymers crystallize with difficulty, because the thermal motion of the monomer units disturb the established order and the crystal breaks. To crystallize a polymer, it is cooled to a temperature where the motion of its units does not hinder their orientation but excessive cooling may also hinder the rearrangement of the units. Crystallization of each polymer is possible only within a definite temperature interval, *i.e.*, between the glass temperature and flow temperature. The glass temperature of isotactic polystyrene is about 100°C. Hence, it will only crystallize between 100°–220°C which is its melt temperature.

Packing of Molecules

Polymers obey the principle of close packing and macromolecules must be packed as closely as possible in the crystalline lattice. There are several possibilities for the formation of close packings of polymer chains.

The first crystalline structure is built by close packing of spheres which is exhibited by globular proteins. The formation of such crystalline structures are possible because all the spheres are equal in size, *e.g.*, edestin catalase, iodobenezoyl glycogen.

The second possibility of close packing is the packing of helical macromolecules. In it convexities of one helix fit into the concavity of the other, *e.g.*, amylose, poly (hexamethylene adipamide), polypeptides (Fig. 3.35a & b).

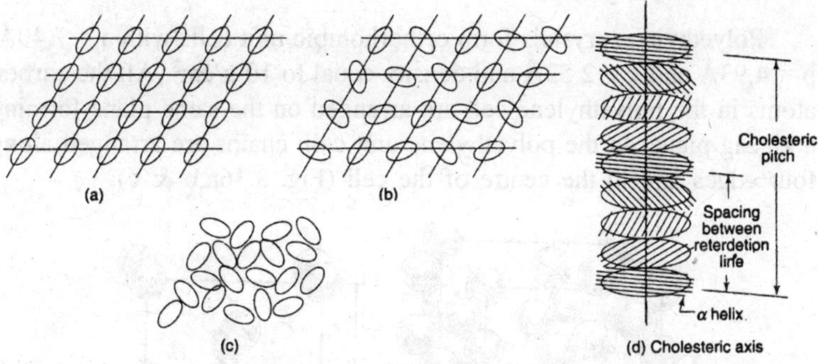

Fig. 3.35 : Different modes of packing of chain molecules (ovals denote projection of monomeric unit of polymer chain): a–long-range order of arrangement of both chains and units in three dimensions; b—ordered arrangement of centres of unit sections and unordered orientation of sections; chains axes form regular lattice; units are arranged disorderly; c –"liquid" arrangement of chain (long-range order of arrangement of units along each chain and short-range order of chain arrangement).

The third is the packing of long extended chains, *e.g.*, polyamide, polyurethanes, etc. The essential requirement is that the side chain substituents must not hinder regular arrangement of adjacent chains.

Close packing of chain molecules can only be accomplished with flexible chains capable of shifting in parts. Introduction of polar groups gives rise to two opposite effects. On one hand intramolecular attraction increases, favouring close packing, while on the other hand, the thermodynamic flexibility (potential difference between two position of the chain) decreases. The chain becomes more rigid. The rate of close packing depends on these two factors. However, with polar groups like –OH or –CN, the attraction effect prevails inspite of the fact that the chains are rigid. Since rigid chains closely pack in the extended state, such polymers are highly oriented, *e.g.*, poly (vinylalcohol), polyacrylonitrile. The chains of isotactic polypropylene are highly flexible and form crystalline lattice, but rigid cellulose chains can- not form such lattices under any condition.

Polymer Crystal Cells

The crystalline cell of polymers consists of parts of chains located at definite distances a, b and c and angles α, β and γ. They do not differ at all from cells formed by low molecular weight compounds and obey all the rules of symmetry.

Polyethylene crystals form orthorhombic unit cells with a = 7.40Å, b = 4.93Å and c = 2.53Å and density equal to 10 g/cm^3. All the carbon atoms in the polyethylene cell are arranged on the same plane forming a zigzag plane. In the polyethylene unit cell, chains are arranged along four edges and in the centre of the cell (Fig. 3.36a,b & c).

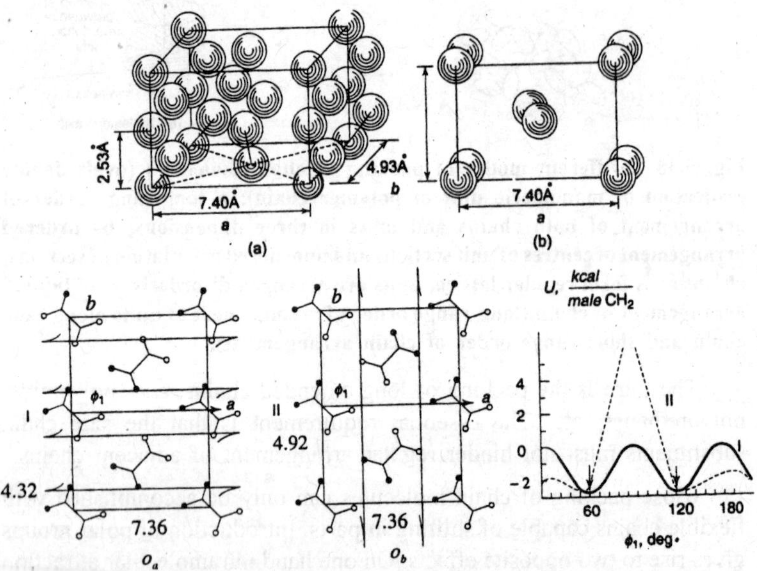

Fig. 3.36 : Conformation of a chain (a) and the arrangement of chains in a crystalline cell of polyethylene (b)

In the case of chains having bulky side substitutents helical crystals are found, *e.g.*, polypeptides, polypropylene, vinyl and olefin polymers (Fig. 3.37). One helical loop may contain a different number of monomeric units depending on the nature of a polymer.

Natta synthesized 1, 2-poly butadiene in the crystalline state with high melting point and excellent mechanical properties. The chains were almost planar and they hold a zigzag structure. The vinyl group substituents were arranged in the 1–3 sequence on both sides of the plane. Similar results were experienced with isotactic polystyrene.

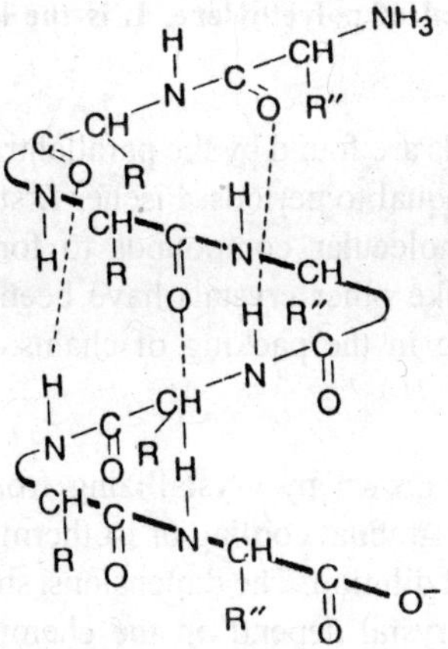

Fig. 3.37 : Right-handed helix. Dashed lines indicate hydrogen bonds of (he large period

Polymers of sulphur or phosphorus are characterized by polymorphism. Sulphur may form crystals belonging to monoclinic, hexagonal and triclinic types of symmetry. They are typical first order phase transitions.

These transitions between various polymorphous formations take place either upon a change in temperature or under the action of load which brings a change in the parameters of a crystalline cell. As a matter of fact most of the polymers can be crystallized under stretched conditions.

The morphology of a crystalline polymer is determined by the mutual arrangement of the unit cells within the limits of the crystalline state of a body.

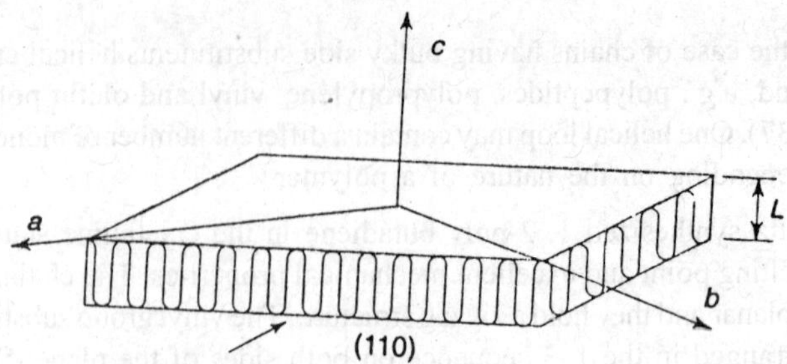

Fig. 3.38 : Diagram of the arrangement of crystallographic axes in a lamellar crystal of polyethylene. L is the length.

Single Crystals

Single or ideal crystals are found by the parallel transfer of unit cells along edges at distances equal to periods. Fischer first of all ascertained the possibility of high molecular compounds to form mono-crystals. Polymer single crystals like other crystals have been found to contain certain defects originating in the packing of chains.

Lamellar Crystals

Polymer crystals are grown by crystallizing from a one per cent solution of polymer under gradual cooling or isothermal aging under the equilibrium temperature of dilution. The dimensions, shape and regularity of structure of a single crystal depend on the chemical structure of a chain and the physical constraints like temperature, concentration of solution, the cooling rate, the nature of the solution, etc. The polymer, single crystals are monolayer flat plane lamella which are often rhomboid of 100Å thick and the size of the plate up to 1 mμ. The axes a and b of a crystalline cell correspond to long and short diagonals of the rhomboid, while, axis c along which the polymer chains are directed is perpendicular to the crystal plane (Figs 3.36 and 3.39). Since the length of the polymer is very big, while thickness of the crystal is of the order of about 100, the long chain can only fit in the crystal by turning on its surface by 180°, This is known as chain fold conformation. To have an idea in a crystal of 120Å thickness, the fold shall contain 100 carbon atoms.

A molecule having a weight of 10^5 shall fold about 70 times.

The thickness of a lamella, the dimensions of crystallite and degree of defects of the boundary layers of a crystal are affected by the crystallization conditions. Crystals with folded chain conformation can be obtained where supercooling is in the range 15Å–20°C.

Crystallization under high hydrostatic pressure $5^{-10} \times 10^3$ atm give extended chain crystal Polyethylene annealed under a pressure of 7000 atm give extended chain crystals. When the melt is cooled under great deformation crystals of different morphological form can be obtained. Polyhedrons known as hedrites and axialites or ovoids are obtained by raising the concentration of the solutions.

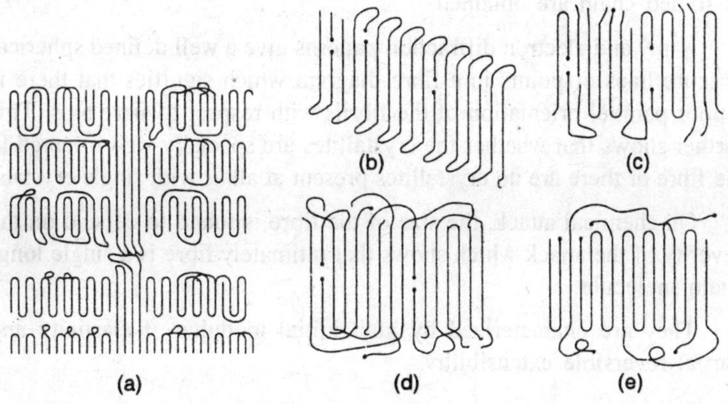

Fig, 3.39 : Two-dimensional schematic illustration of folding surfaces in polymer crystals; a-e-possible cases of the arrangement of chains and their ends on folding surfaces

FIBRIL

Fibrillar crystals are obtained by the high rate of evaporation of a solvent from a moderately concentrated solution or by the cooling of melt.

The crystallization of polymers can be greatly speeded up by stretching or drawing the solution or the polymer melt during or after cooling. This is carried out by pressing the melt or the solution through spinerets. Crystallization occurs during this process and the threads take up a typical fibre structure. At first a crystalline structure is formed in which the molecular chains are perpendicular to the axis, but by stretching on cooling a reorientation of the polymer chains parallel to the fibril take place. During the process of drawing, a parallel orientation of the polymer chains and the chain segments takes place until the chains are so strongly interlocked because of crystallization that further gliding a part is impossible, *e.g.*, polyamides.

There are two views about their formation. One section believes that aggregation of rolled lamella is responsible for them while others hold that lamella degenerate in the formation of fibrillar crystals.

In stereospecific polymerization, fibrillar crystals are directly formed. The formation is determined by the ratio of the rates of growth of a chain and its crystallization. By changing this ratio, crystal either with extended or folded chain are obtained.

X-ray and electron diffraction patterns give a well defined spherical obscure lines or points in a fibre diagram which signifies that there is a poor parallel orientation of the fibrils with respect to fibre axis. This further shows that whether the crystallites are specially well oriented in the fibre or there are no crystallites present at all as with single crystals.

On chemical attack, the size of the fibre is found to depend on the severity of the attack which shows that ultimately fibre is a single long-chain molecule.

They are characterized by high initial modulus of elasticity and partial reversible extensibility.

FILM OR SHEETS

The term film is used for membranes usually 0.15 mm thick, while sheets are thicker than this. As already explained in the case of fibres, if instead of spinerets, a slit is used for stretching the melt or the solution, a film or sheet results depending upon the width of the slit. Biaxially oriented films have high strength and low water vapour transmission. The biaxial orientation depending upon the stretching can be equal or unbalanced in the two direction. They are usually transparent or translucent.

SPHERULITES

Spherulites are spherical crystallites ranging from submicroscopic size to 100 microns in diameter. They represent the crystalline portion of the sample, growing at the expense of non-crystalline melt. In its formation, the nucleus may be a foreign particle or it may arise spontaneously. They are obtained when polymers crystallize from melts or concentrated solutions. Polarized light reveals that the polymer chains are oriented tangentially around each nucleus.

Macromolecular chains make at least an angle of 60° with the radius, *i.e.*, they are arranged tangentially to the spherulite radius. Spherulites possess anisotropic properties.

Their indices of refraction of light are different in radial or tangential directions. The orientation of the crystallographic axes continuously changes along the angular coordinates. A spherulite is called positive when the refractive index in the radial direction is greater than in the tangential one, otherwise it would be a negative spherulite.

In radial Spherulites one of the axes of the crystalline lattice retains a constant direction along all the radial directions. Ring type Spherulites are built of lamellar crystals whose orientation continuously changes along the spherulite radius.

Depending on the degree of cooling, the same polymer may form different types of Spherulites. With low degrees of supercooling Ring type of Spherulites are obtained while radial Spherulites are obtained by high degree of supercooling.

The formation of Spherulites passes through several stages. First the crystal nucleus is formed which are scattered throughout the volume of the sample and then independent growth of discrete fibrillar or lamellar crystalline structure grow at the same time in all directions. Then encountering one another they grow into regions filling the entire volume of the body and their boundaries are disturbed, they assume the form of a polyhedron.

The structural units in spherulite are bundles of macromolecules made out of chains in an extended conformation. This is possible when several macromolecules try to crystallize simultaneously from both the ends. Such interstructural bonds combine separate crystallites inside fibrillar or lamellar crystals.

Sometimes with the periodic twisting of the lamellae, a ring structure is obtained in Spherulites.

Because of the lamellar structure, they display a circular birefringence area possessing a dark cross pattern when viewed in polarizing microscope.

If a crystallite is stressed beyond the elastic limit, rearrangement of the polymer chain occurs, in which chains are oriented in the direction of the stress, destroying the spherulite pattern. This technique is largely employed in fibres.

GLOBULAR CRYSTALS

In globular crystals, macromolecules in coiled conformation form the lattice points. These are found in biopolymers as a very high degree of uniformity of macromolecules is a prerequisite for it, *i.e.*, tobacco mosaic virus. Synthetic polymers do not form them.

SHAPE OF THE MACROMOLECULE

The shape of the macromolecule can be determined by the special arrangement of atoms and the constitution of bonds. Macromolecules have a chain structure. At present we will not consider structure where branching or crosslinking of substituents occurs that hinder the free rotation, because the problem of molecular shape will become more difficult. The paraffin chain can assume different shapes as a result of rotation around the C—C axis. The chain molecules will always try to assume a structure of maximum possible entropy, *i.e.*, it will assume most probably a shape which is most irregular. It can be statistically proved that in a chain of 4000 atoms, 10 chain molecule models will be obtained out of which only one would be a fully stretched chain. In case of two molecules A & B, if we try to establish a regularity between them by throwing the dice a large number of times, it is quite rare. As a matter of fact by throwing the dice 1000 times one obtains the shape of a long chain as a statistical coil.

CRAFT COPOLYMER

A polymer with branched molecules, in which the main chain is chemically different from the branches, is known as graft copolymer. A monomer after polymerization, is subsequently subject to the influence of another monomer which polymerizes into the primary high polymer chain, results in a graft copolymer (Fig. 3.40).

```
—AAA—AAA—AAA—
  |     |      |
  B     B    B
  |     |      |
  B     B    B
```

The graft copolymers have properties combining some of the properties of the two polymers. It has become possible to attach immiscible polymers to each other in them, but the molecules of the two components are spatially separated from each other. The graft copolymers are prepared either by means of radicals or functional groups or copolymerization. Vinyl acetate was polymerized in presence of styrene vinylidene chloride copolymer. Some of the vinyl acetate radicals abstracted chlorine from the polymer chain, after which vinyl acetate branches were produced. The resulting graft copolymer had a hydrophobic backbone and hydrophilic branch. Similarly, polyethylene can be irradiated with accelerated electrons which leaves peroxide free radicals trapped on the polymer backbone. If this is exposed to acrylonitrile, branches of polyacrylonitrile grow on the polyethylene system at the free radical sites.

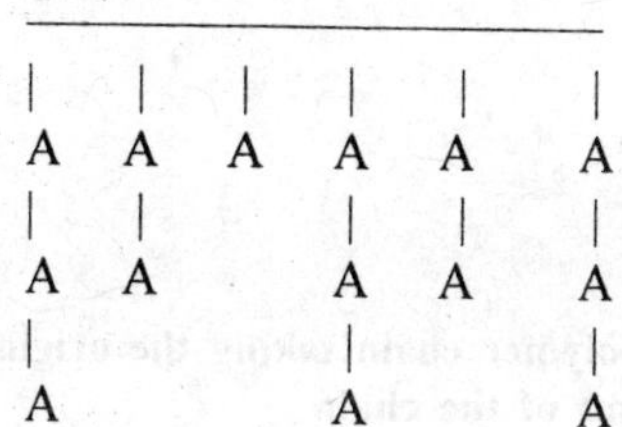

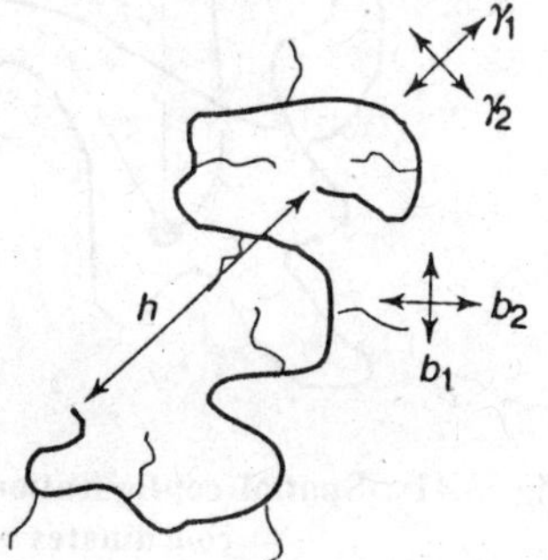

Fig 3.40

RANDOM WALK MODEL

Kuhn and Maik were the first to demonstrate that the macromolecules exist as coils. This can be explained by random walk model. Let r be the distance from one end group to the other of the chain molecule. This is also known as displacement length. The length of the fully extended chain is called as contour length. The average value of r is taken as $(r^2)^{1/2}$. The effective size of the polymer molecule is given by the root mean square distance of the elements of the chain from its centre of gravity. This is also known as radius of gyration represented by $\sqrt{S^2}$.

A chain consisting of n vectors of equal magnitude but unrestricted direction will give a random model (Fig. 3.41). Let it consist of l linkages.

The average square of the projection of a bond vector on x-axis is given by—

$$l_x^2 = \int_0^l l_x^2 \, p(l_x) \, dl_x = \frac{l^2}{3} \qquad ...(49)$$

where p (l_x) dl_x is the probability that the projection lies between l_x and l_x + dl_x.

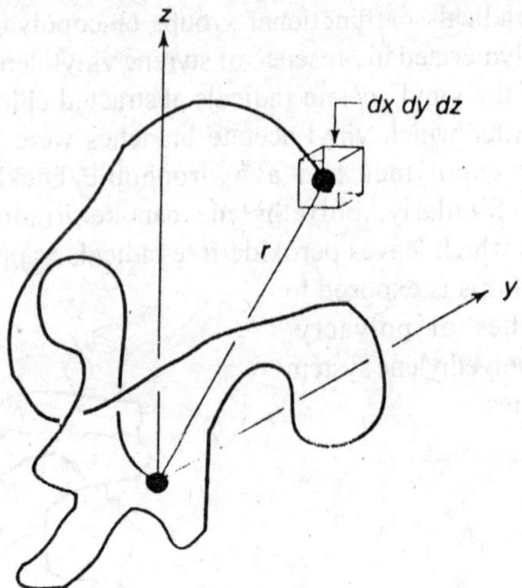

Fig. 3.41 : Spatial configuration of a polymer chain taking the origin of coordinates at one end of the chain

$$\left(\bar{l}_x^2\right)^{1/2} = \frac{1}{\sqrt{3}} \qquad \text{...(50)}$$

If r is the magnitude of the chain displacement Vector r

$$r^2 = x^2 + y^2 + z^2$$

The total volume of all volume elements at a distance r from the origin is 4 π r² dr. The probability that the chain displacement length has a value in the range r and r + dr

$$W(r)\,dr = \left(\frac{\sqrt{3}}{2} \Big/ n^{1/2}\, l\, \pi^{1/2}\right)^3 \exp\left(\frac{3r^2}{2l^2 \times n}\right) 4\pi r^2 dr \qquad \text{...(51)}$$

$$\bar{r} = \int^{\alpha} rW(r)\,dr \Big/ \int^{\alpha} W(r)\,dr \qquad \text{...(52)}$$

$$= \sqrt{\frac{8}{3\pi}}\, l n^{1/2}$$

Average square end to end length is $r^2 = \int_0 r^2 W(r)\,dr$ ...(53)

$$\sqrt{\bar{r}^2} = l n^{1/2} \qquad \text{...(54)}$$

and $\sqrt{\bar{S}^2} = \sqrt{\bar{r}^2/6}$ ...(55)

Figures 3.42(a) and (b) give the variation of gaussian density and W(r) with the increasing value of r.

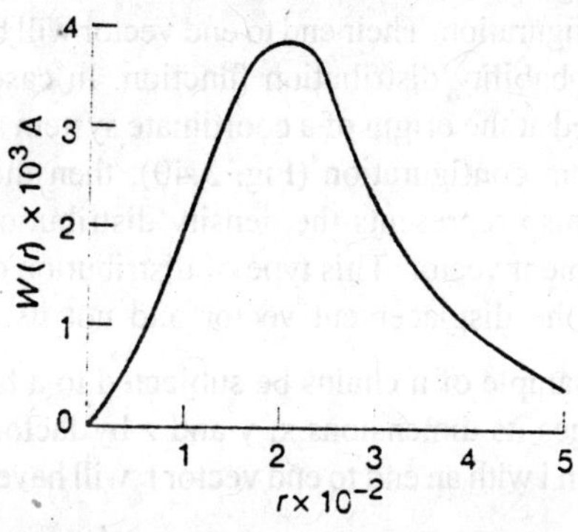

(a)

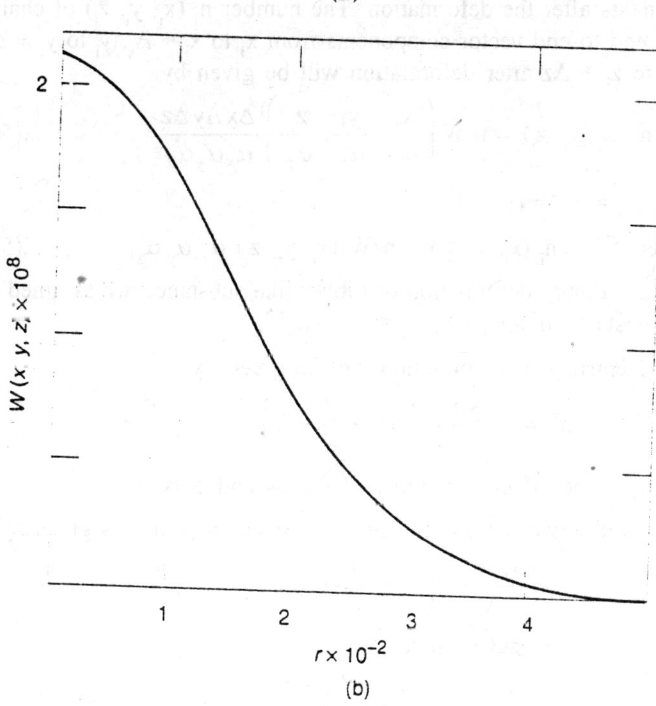

(b)

Fig. 3.42

Random Theory

Let us consider a network of n chains and also assume that the actual cross-linking process is carried out in isotropic (unoriented) polymers. Since the cross-linking is a random process, so the created chains will also be of random configuration. Their end to end vector will be distributed according to their probability distribution function. In case one end of polymer chain is placed at the origin of a coordinate system and the chain is assumed any random configuration (Fig. 3.40), then the probability distribution function also represents the density distribution of the end points of the displacement vector. This type of distribution depends only on the magnitude of the displacement vector and not its direction.

Let the network sample of n chains be subjected to a homogeneous strain, so that it changes its dimensions x, y and z by factors α_x, α_y and α_z respectively. A chain i with an end to end vector r_i will have components $\frac{x_i}{\alpha_x}, \frac{y_i}{\alpha_y}$ and $\frac{z_i}{\alpha_z}$ before the deformation when x_i, y_i and z_i are its components after the deformation. The number n_i (x_i, y_i, z_i) of chains having end to end vector components from x_i to $x_i + \Delta_x$, y_i to $y_i + \Delta y$ and z_i to $z_i + \Delta z$ after deformation will be given by

$$n_i\ (x_i, y_i, z_i) = n\ W\left(\frac{x_i}{\alpha_x}, \frac{y_i}{\alpha_y}, \frac{z_i}{\alpha_z}\right)\frac{\Delta x\,\Delta y\,\Delta z}{\alpha_x\alpha_y\alpha_z} \qquad ...(56)$$

If $\alpha_x = \alpha_y = \alpha_z = 1$

then $$n_i\ (x_i, y_i, z_i) = n\ W\ (x_i, y_i, z_i)\ \alpha_x\ \alpha_y\ \alpha_z \qquad ...(57)$$

The ordinary deformation of rubber-like substance are assumed to be at constant volume, *i.e.*, $\alpha_y = \alpha_z = \alpha_z^{1/2}$.

The entropy of deformation will be given by

$$\Delta S = \left(-\frac{Kn_e}{2}\right)\left(\alpha^2 + \frac{2}{\alpha} - 3\right)$$

where n_e is the effective number of chains and α is α_x

The retractive force T per unit cross sectional area is given by

$$\tau = \left(\frac{RT\,n_e}{V}\right)\left(\alpha - \frac{1}{\alpha^2}\right)$$

where n_e is expressed in moles

$$= \left(RT\frac{n}{V}\right)\left(1-\frac{2N}{n}\right)\left(a-\frac{1}{\alpha^2}\right)$$

where V is the volume of the deformed chain and N represents the number of primary molecules amongst which the cross linkages occur.

ELASTOMERS

Polymers which at the room temperature or below undergo deformation under mechanical stress are called elastomers. It involves the entropy effect and the structural elements become oriented so that the system under stress tries to approach crystallinity. The release of stress increases the entropy by allowing the chains to assume more probable shapes.

Elastomers have covalently crosslinked micromolecules as a consequence of which two conditions arise.

1. The number of cross-linking points are independent of temperature.
2. The position of cross-linking points is not changeable.

Thus, it is possible to produce soft rubber with high elasticity. The reason that elastomers are highly elastic is that the number and position of cross-linking points are fixed which prevents a spatial arrangement of the macromolecules with respect to each other. The coils are not able to pass each other, so they remain in their stretched form for longer time. Such covalent linking is carried out by a process known as vulcanization. The higher the cross-linking, the larger would be the modulus of elasticity.

In the case of natural rubber and butadiene styrene copolymers, the vulcanization occurs by the reaction of sulphur on the double bond, but in thc case of saturated polymers, *e.g.*, ethylene propylene copolymor, or polyisobutylene, which are also elastic, the room temperature vulcanization is carried out by building certain suitable reactive groups.

Ferry and others have discussed in detail about the elastic state of polymers and have studied the effect of temperature on rubber. The following observations were obtained:

1. The glass state begins at –80° for natural rubber. The molecular movement stops at this stage and the rubber becomes hard and brittle. The elasticity depends on the potential energy of the molecules.

2. The transition temperature T is not well defined as it depends on the interaction between the chains. The modulus shows a sharp. decrease (Fig. 3.43).

The following properties were observed.

Table 3.14

Properties	*Crystalline State*	*Glassy State*
1. Influence of temperature	Takes place over a broad temperature range. An optimum temperature exist for each polymer.	Occurs over a small temperature range.
2. Appearance	Colour of raw natural rubber changes from brown to opaque grey. Vulcanized rubber becomes stiff but not brittle.	Rubber becomes brittle and hard like glass.
3. Latent heat	Heat is developed during crystallization.	No heat affect, sp heat changes.
4. Speed of transition	Shorter or longer time is required depending on pressure and the rate of change of temperature.	Transition usually rapid.
5. Temp. Volume relationship	Decrease of volume.	Change of the coefficient of expansion.

Glasses corresponding to start type of liquid exhibit strongly anisodimensional molecules, *e.g.*, vitreous sulphur, phospho-chloronitrides. Tamman1 and Bastress have prepared rubber-like materials from hydrated polyphosphates. Glasses and elastomers are usually quite far removed from the state of an ideal elastic solid. Hoode's law is applicable to ideal solid and Newtonian law to ideal liquids, then the plot between elongation and deformation time can be represented as (Figs 3.44, 3.45) where deformation times are short and loads applied are also small. But it takes longer time to flow, then the curve which is shown in Fig. 10.13a and their continuation is given in Fig. 10.13b.

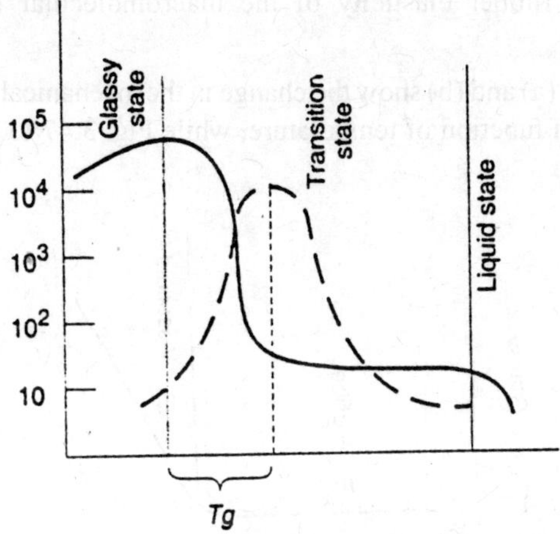

Fig. 3.43

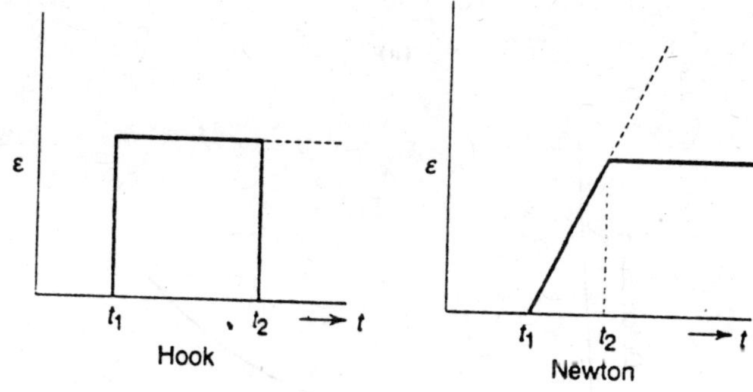

Fig. 3.44 Fig. 3.45

Maxwell obtained a relation

$$\sigma = \varepsilon_0 \text{ E exp} - \left(\frac{E}{\eta}\right) t = \varepsilon_0 \text{ E exp}\left(-\frac{t}{\tau}\right)$$

where ε is the elongation, σ, the force per cm^2 and E, the modulus of elasticity, t, the time and η, the viscosity.

The η/τ is called the relaxation time t. Therefore, the smaller the relaxation time, the larger the modulus of elasticity at a given viscosity. The polymer becomes highly viscous in the region of softening. On further heating, rubber elasticity of the macromolecular compound appears.

Figures 3.46 (a) and (b) show the change in the mechanical behaviour of a polymer as a function of temperature, while Fig. 3.47 to 3.50 show deformation.

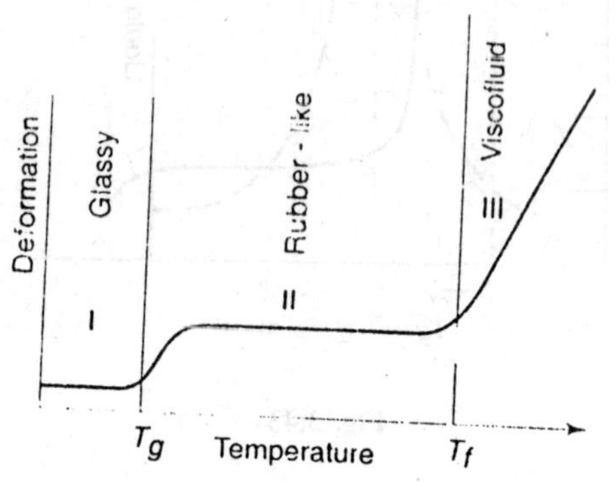

(a)

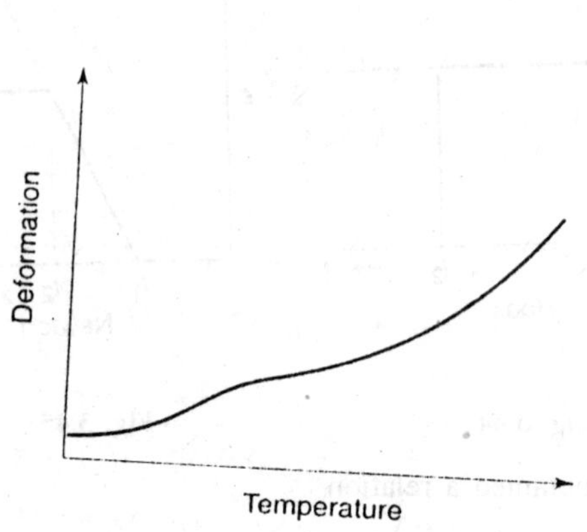

(b)

Fig. 3.46 : Thermodynamical curve of a polymer.

The rubber elastic deformation involves molecular groups and molecular segments within one macromolecule and not the coils themselves. The intramolecular flow process stops by itself because of the deformation of the statistical coil. Lamm has obtained a relation between viscosity and their degree of polymerization.

The changes in the relaxation phenomena with macromolecular compounds are determined by their structures, *e.g.*, polystyrene has only one damping maximum. If the addition of rubber is carried out in such a way that the phase boundaries are tied to each other by grafting, industrially important high impact polystyrene can be prepared.

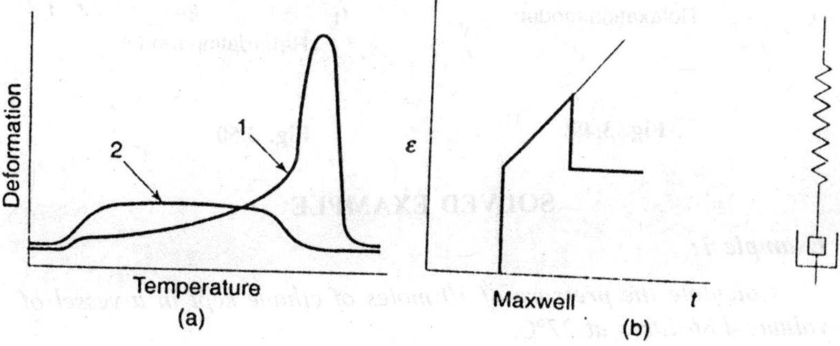

Fig. 3.47

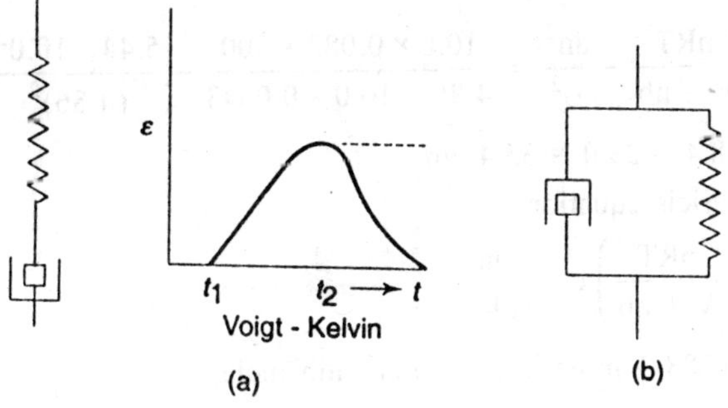

Fig. 3.48

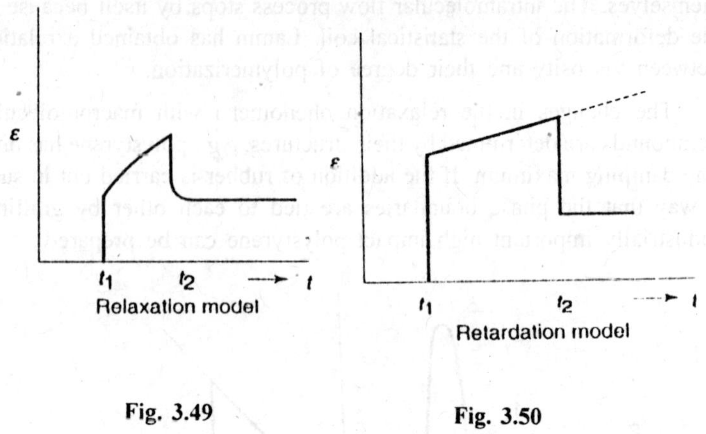

Fig. 3.49 Fig. 3.50

SOLVED EXAMPLE

Example 1:

Calculate the pressure of 10 moles of ethane kept in a vessel of volume 4.86 litres at 27°C.

Solution:

(a) Ideal gas equation

$$P = \frac{nRT}{V} = \frac{100 \times 0.082 \times 300}{4.86} = 50.7 \text{ atm}$$

(b) Van der Waals equation

$$P = \frac{nRT}{\upsilon - nb} - \frac{an^2}{\upsilon^2} = \frac{10.0 \times 0.082 \times 300}{4.86 - 10.0 \times 0.0643} - \frac{5.44 \times 10.0^2}{(4.86)^2}$$

$= 58.4 - 23.0 = 35.4$ atm.

(c) Dietericis equation

$$P = \left(\frac{nRT}{V - nb}\right) e - \frac{an}{VRT}$$

$b = 0.708 \ 1 \text{ mole}^{-1}$; $a = 7.111^2 \text{ atm mole}^{-2}$

$$P = \left(\frac{nRT}{V - nb}\right) e - \frac{an}{VRT} \quad e = 7.11 \times 10/4.86 \times 0.082 \times 300$$

$= 59.3 \ e^{-0\text{-}594} = 32.7$

(d) Reduced equation

$$T_R = \frac{T}{T_c} = \frac{300}{191} = 157$$

$$V_r = \frac{V}{V_c} = \frac{4.86}{0.99} = 4.9$$

$$P_r = \frac{P}{P_c} = \frac{P}{45.8}$$

From Eq. 126

$$P = P_c P_r = 45.8 \times .71 = 32.6 \text{ atm}$$

It can also be obtained by finding out the value of P_r from the value of β for methane in Fig. 3.51 but the actual value observed by B.H. Sage *et al.* was 34.0 atm

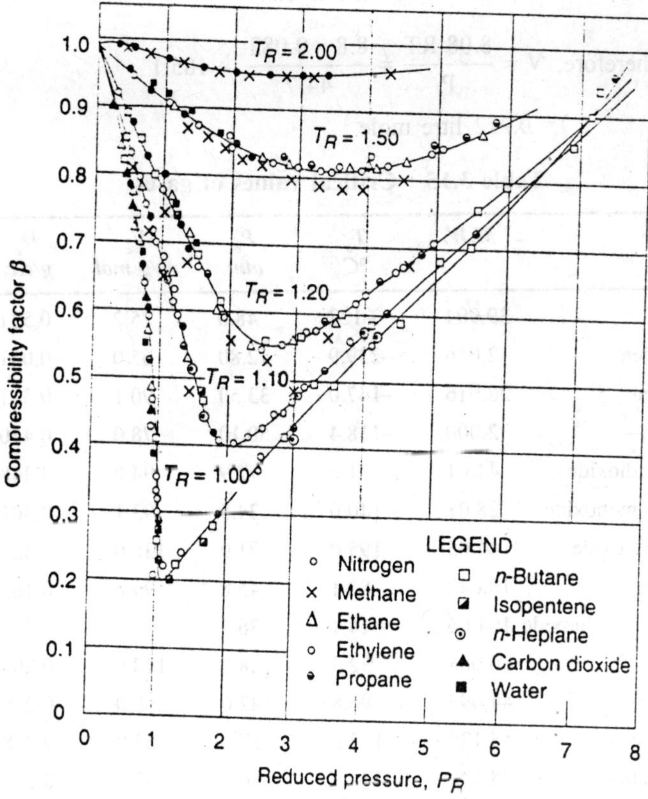

Fig. 3.51

Thus, marked deviations are observed in almost all the cases showing that all these including the corresponding states relations are only approximations. However, the corresponding states relations have proved a very quick method of finding out the parameters of a non-ideal gas.

Example 2:

Estimate the volume of one mole of oxygen at –88°C and 44.7 atm.

$$\frac{T}{T_c} = \frac{273-88.0}{154.4} = 1.2$$

$$\frac{P}{P_c} = \frac{44.7}{49.7} = 0.9$$

Solution:

The value of β from Fig. 3.51 is 0.8.

$$\text{Therefore, } V = \frac{8.08\ RT}{P} = \frac{8.8 \times 0.082}{44.7} \times 185.1$$

$$= 0.272 \text{ litre mole}^{-1}$$

Table 3.15 : Critical values of gases

Gas	*M.W.*	T_c °C	P_c atm	V_c cc/g mol	D_c g/c.c.
Argon	39.994	–122	48.0	75.2	0.531
Hydrogen	2.016	–239.9	12.80	65.0	0.031
Nitrogen	28.016	–147.0	33.50	90.1	0.311
Oxygen	32.000	–118.4	50.10	78.0	0.410
Carbon dioxide	44.01	31.0	72.9	94.0	0.468
Carbon monoxide	28.01	–140.0	34.5	93.1	0.301
Ethylene oxide	44.05	195.0	71.0	138.0	0.320
Methane	16.042	–82.1	45.8	99.0	0.162
Silicon tetrafluoxide	104.06	–14.1	36.7	–	–
Ethane	30.068	32.3	48.2	148.0	0.203
Propane	44.094	96.8	42.0	200.0	0.220
n-Butane	58.120	152.0	37.5	255.0	0.228
Iso-Butane	58.120	134.9	36.0	263.0	0.221
Pentane	72.146	196.6	33.3	311.0	0.232

Iso-pentane	72.146	187.8	32.9	308.0	0.239
Neo-pentane	72.146	160.6	31.6	303.0	0.238
Ethylene	28.052	9.9	50.5	124.0	0.227
Propylene	42.07	91.8	45.6	181.0	0.233
1-Butene	56.104	146.4	39.7	240'.0	0.234
2-Butene	56.104	157.0	41.0	236.0	0.238
iso-Butene	56.104	144.7	39.5	239.0	0.235
1-Pentene	70.130	201.0	40.0	–	–
2-Pentene	70.130	203.4	40.4	–	–
Propadiene	40.062	120.0	–	–	–
1-3 Butadiene	54.088	152.0	42.7	221.0	0.245
Propyne (ally line)	40.062	128.0	52.8	–	–
Ethyl Benzene	106.16	346.4	38.0	–	–
Silicon Tetrachloride	169.89	233.0	–	–	–

Example 3:

A sample of oxygen is collected by displacing water from an inverted tube. The temperature is 25°C, the pressure is 750 mm Hg (barometric pressure) and the volume occupied is 280 cm³. What is the true volume of oxygen at STP?

Solution:

As the gas is collected over water barometer pressure (P_{bar}) does not correspond to the actual pressure of the gas

$$P_{bar} = + PH_2O$$

$$P_{O_2} = P_{bar} - P_{H_2O}$$

$$= (750.0 - 23.8) \text{ mm Hg} = 726.2 \text{ mm Hg}.$$

Volume of gas at STP can be calculated as:

Initial state: $P_1 = 726.2$ mm Hg; $V_1 = 280$ cm³;

$n_1 = n$; $T_1 = 273 + 25 = 298°K$

$P_1V_1 = nRT_1$

Final state: $P_1 = 760$ mm Hg, $V_2 = ?$ $n_2 = n$; $T_2 = 273$ $P_2V_2 = nRT_2$

As nR is the same in the two states, therefore,

$$\frac{P_1V_1}{T_1} = \frac{P_2V_2}{T_2}$$

$$V_2 = \frac{P_1}{P_2} \times \frac{P_2V_2}{T_2}$$

$$= \frac{726.2 \text{ mm Hg}}{760 \text{ mm Hg}} \times \frac{273° \text{ K}}{298° \text{ K}} \times 280 \text{ cm}^3$$

$$= 245 \text{ cm}^3.$$

Example 4:

You are given three vessels A, B and C, all at the same temperature, with volumes respectively of 1.20 1, 2.63 1 and 3.05 1. Vessel A contains 0.695 g of nitrogen gas at a pressure of 742 mm Hg, vessel B contains 1.10 g argon gas at a pressure of 383 mm Hg, vessel C is completely empty at the start of the experiment. What will the pressure become in vessel C if the contents of A and B are completely transferred to C?

Solution:

Nitrogen was originally present in vessel A (V = 1.2 1) and has a pressure of 742 mm Hg. It is now transferred to vessel C, the volume of which is 3.05 1.

Argon was originally in vessel B (V = 2.63 1) and has a pressure of 383 mm Hg. This is transferred to vessel C(C = 3.05 1).

Now we can treat the two gases separately and find their pressures when present in vessel C. The total pressure in vessel C is equal to the sum of the pressures of nitrogen and oxygen in vessel C.

Gas	*Vessel*	*Initial conditions*	*Final conditions*	*Partial pressures (P) in vessel C*
				$P_1V_1 = P_2V_2$
Nitrogen	AC	$V_1 = 1.2$ 1	$V_2 = 3.05$ 1	$P_{N_2} = \frac{1.21 \times 742 \text{ mm Hg}}{3.05 l}$
		$P_1 = 742$ mm Hg	$P_2 = ?$	= 294 mm Hg
Argon	BC	$V_1 = 2.631$	$V_2 = 3.05$ 1	$P_{Ar} = \frac{2.631 \times 383 \text{ mm}}{3.051}$
		$P_1 = 383$ Hg	$P_2 = ??$	

Pressure of gaseous mixture in C

$= PN_2 + P_{Ar}$

$= 294$ mm Hg + 330 mm Hg

$= 624$ mm Hg.

Example 5:

(a) *What is the average volume available to a molecule in a sample of nitrogen gas at STP?*

(b) *Assuming that N_2 molecule is approximately spherical with an effective diameter of 3.6Å, calculate*

(i) *The actual volume occupied by a N_2 molecule;*

(ii) *the percentage of the molar volume that is empty space;*

(iii) *The average distance between neighbouring gas molecules at STP.*

Solution:

We know that one mole of an ideal gas at STP (*i.e.*, at 0°C and 1 atm pressure) occupies 22.4 litres.

Therefore, assuming ideal behaviour for nitrogen, one mole of this gas will occupy 22.4 litres or 22,400 cm^3 at 0°C and 1 atm, pressure.

There are 6.02×10^{23} molecules in one mole. Therefore, the number of molecules in 1 cubic centimeter is

$$\frac{6.02 \times 10^{23}}{22,400} = 2.69 \times 10^{19} \text{ molecules per cm}^3$$

and the volume available to a gas molecule is

$$\frac{1}{2.69 \times 10^{19}} = 3.72 \times 10^{-20} \text{ cm}^3 \text{ molecule}^{-1}$$

(b) Volume of a sphere $= \frac{4}{3}\pi r^3$

As diameter for nitrogen is 3.6 Å $= 36 \times 10^{-8}$ cm

$$r = \text{radius of nitrogen molecule} = \frac{3.6 \times 10^{-8}}{2} \text{cm}$$

$$= 1.8 \times 10^{-8} \text{ cm.}$$

$$\text{Volume of a nitrogen molecule} = \frac{4}{3}(1.8 \times 10^{-8})^3$$

$$= 2.44 \times 10^{-23} \text{ cm}^3 \text{ molecule}^{-1}$$

Volume of 1 mole of nitrogen molecules

$= 2.44 \times 10^{-23} \times 6.02 \times 1023 = 14.7 \text{ cm}^3 \text{ mole}^{-1}$

We find that the actual volume occupied by 1 mole of molecules is 14.7 cm^3 while the gas molar volume is 22,400 cm^3. Thus, the difference 22,400 – 14.7 = 223,85.3 cm^3 is apparently the empty space.

$$\text{per cent empty space} = \frac{\text{Volume empty}}{\text{Volume available}} \times 100$$

$$= \frac{223,85.3}{224.00} \times 100 = 99.9\%$$

The average distance between any molecule and its nearest neighbour can be obtained from the volume available to a gas molecule: Let us assume that the volume available to a molecule has the shape of sphere of radius R. Hence, the volume available to a molecule is 4/3 R^3 and this is equal to 3.72×10^{-20} cm^3 as found in (a)

$$4/3\ R^3 = 3.72 \times 10^{-20} \text{ cm}^3$$

$$R^3 = \frac{3 \times 3.72 \times 10^{-20}}{4x}$$

$$R = 20.7 \times 10^{-8} \text{ cm}$$

Thus, on an average, the volume available to a gas molecule can be assumed to be the volume of a sphere with R = 20.7×10^{-8} cm^3, and accordingly the average distance between the centres of any two neighbouring molecules in the gas sample will be 2R.

Therefore, the average distance between a gas molecule and its nearest neighbour is 2R = $20.7 \times 10^{-8} = 41.4 \times 10^{-8}$cm = 41.4Å.

Another way of computing the distance between a gas molecule and its nearest neighbour in a sample of gas is by assuming that the gas volume is divided into small cubes of edge-length and the volume of this small cube is the volume, on an average, available to a gas molecule. The length of this small cube (1) will then roughly be equal to the average distance between the gas molecules assuming that the molecules are for most part evenly spaced at the centres of identical cubes. If 1^3 is the volume of each small cube, then the value of 1 can be computed from the relation

$$1^3 = 3.72 \times 10^{-20} \text{cm}^3$$

where 3.72×10^{-20} is the volume available to each molecule or l = 33.4 × 10^{-8} cm = 33.4Å

From these calculations, we find that the average distance between two neighbouring molecules at STP is about 33.4 × 10^{-8} cm or 33.4Å (students should find out why this value is different from that obtained previously). The diameter of the gas molecule is only 3.6Å and the distance between molecules in the gas is, therefore, about ten times their diameter.

Similar calculations for solid (density 1.03 gm/cm^3) and liquid (density 0.8 gm/cm^3) nitrogen show that in these states, the average distance between molecules is about 3.56Å and 3.8Å respectively (which are nearly equal to the molecule diameter of nitrogen).

Hence, in solid or liquid states, individual molecules touch each other and it is difficult to compress the substances in these states unlike in the gaseous state.

Example 6:

Naphthalene on oxidation gives phthalic anhydride, a very important raw material for glyptal type of resins. When it is heated to 30°C, it exerts a vapour pressure of 0.177 mm. Calculate the effusion rate when M = 128.16 g/g.mole.

Solution:

0.177 mm = 2.36 × 10^5 g-cm/sec^2

$$Z_m = \frac{\alpha}{4} P \left(\frac{3M}{RT}\right)^{1/2}$$

$$= \frac{0.92}{4} \times 2.36 \times 10^5 \left(\frac{3 \times 128.16}{8.314 \times 303 \times 10^7}\right)^{1/2}$$

$$= 0.23 \times 2.36 \times 10 \left(\frac{128.16}{8.314 \times 10^7}\right)^{1/2}$$

$$= 6.73 \text{ g/cm}^2 \text{ sec.}$$

Example 7:

Compute the relative rates of diffusion of hydrogen and carbon dioxide. Molecular mass of hydrogen is 2 and that of carbondioxide is 44.

Solution:

$$\frac{r_{H_2}}{r_{CO_2}} = \frac{(M_{CO_2})^{1/2}}{(M_{H_2})^{1/2}} = \left(\frac{44}{4}\right)^{1/2}$$

$$= \frac{6.63}{1.41} = 4.7$$

Hydrogen diffuses 3.7 times faster than ,carbon dioxide.

Example 8:

The speeds of diffusion of CO_3 and O_3 were found to be as 0.29 and 0.271. What is the molecular mass of O_3 if the molecular mass of CO_2 is 44?

Solution:

$$\frac{r_{O_3}}{r_{CO_2}} = \left(\frac{M_{CO_2}}{M_{O_3}}\right)^{1/2}$$

$$\frac{0.271}{0.29} = \left(\frac{44}{M_{O_3}}\right)^{1/2}$$

Squaring both sides

$$\frac{(0.271)^2}{(0.29)^2} = \frac{44}{M_{O_3}}$$

or $$M_{O_3} = 44 \times \frac{(0.29)^2}{(0.271)^2} = \frac{44 \times 0.29 \times 0.29}{0.271 \times 0.271}$$

= 50.4 (Mol. mass of ozone is 50.4).

Example 9:

A sample of 1.0 g of gas contained in a 1.0 litre vessel at 24°C exerts a pressure of 0.836 atm. What would be the pressure of the sample if it was compressed to a volume of 0.9631 and heated to a temperature of 47°C.

Solution:

For the gas in the two states we have

Initial state: $P_1 = 0.836$ atm; $V_1 = 1.01$;

$n_1 = N$; $T_1 = 273 + 24 = 297°K$ $P_1V_1 = nRT_1$

Final state: $P_2 = ?;\ V_2 = 0.9631;\ n_2 = n;$

$T_2 = 273 + 47 = 320°K\ P_2V_2 = nRT_2$

In this problem n is the same in the two states, therefore,

$$\frac{P_1V_1}{T_1} = \frac{P_2V_2}{T_2}$$

or $$P_2 = \frac{V_1}{V_2} \times \frac{T_2}{T_1} \times P_1$$

Substituting the numerical values

$$P_2 = \frac{1.01}{0.9631} \times \frac{320°\,K}{320°\,K} \times 0.863 \text{ atm}$$

$$= 0.935 \text{ atm.}$$

Example 10:

A mixture of gases at 760 mm Hg pressure contains 65.0% nitrogen, 15.0% oxygen and 20.0% carbon dioxide by volume. What is the partial pressure of each in mm Hg?

Solution:

A fundamental property of gases is that each component of a gas mixture occupies the entire volume of the mixture. The percentage composition by volume refers to the volume of the separate gases before mixing. Thus, 65 volumes of nitrogen, 15 volumes of oxygen and 20 volumes of carbon dioxide, each at 760 mm Hg pressure are mixed to give 100 volumes of mixture at 760 mm Hg.

For a mixture of ideal gases, Dalton's and Amagat's laws are equivalent. Therefore, in a gaseous mixture both the volume fraction υ_i/υ and the pressure fraction P_i/P of a component equal its mole traction $x_i = n_i/n$.

Now as

$$P_i = \frac{\upsilon_i}{V} P$$

Therefore, $$P_{N_2} = \frac{65}{100} \times 760 \text{ mm Hg} = 494 \text{ Hg}$$

$$P_{O_2} = \frac{15}{100} \times 760 \text{ mm Hg} = 114 \text{ mm Hg}$$

$$P_{CO_2} = \frac{20}{100} \times 760 \text{ mm Hg} = 152 \text{ mm Hg}$$

Total pressure $= p_{N_2} + p_{O_2} + p_{CO_2} = 760$ mm Hg

Dalton's law is put to practical use in laboratory work when the amount of gas evolved in a reaction is to be measured quantitatively. Often these gases are collected over water. It is common practice to equalize the level of water inside and outside the gas collecting bottle to make the pressure of the collected gas equal to the pressure of the atmosphere. This is done in order to measure the volume of gas at atmospheric pressure and temperature. If the gas was pure, one could immediately use the gas law to calculate the number of moles (or the quantity) of gas produced by the reaction. However, under the conditions of the experiment the gas collected contains water vapour in addition to the gas of interest. Hence, the pressure of the collected gas is not the atmospheric pressure (P_{atm}) even though the equality of the liquid levels indicates that the total internal pressure is equal to P_{atm}; for this total pressure, according to Dalton's law is made up of the sum of the partial pressure of the gas in question (P_g) and the vapour pressure exerted by the liquid water (p_1) over which the gas is collected. Therefore, $P_{atm} = P_g + P_1$. Pressure of the water vapour is also known as aqueous tension. The pressure of water vapour is dependant on temperature and its values at different temperatures may be found in the book of tables. Whenever a gas is collected over water the pressure of the gas is obtained by subtracting the aqueous tension from the atmospheric pressure (recorded by barometer).

Example 11:

A given sample of ideal gas occupies a volume of 11.21 at 0.863 atm. If the temperature is kept constant, to what pressure will you have to go to change the volume to 15.01?

Solution:

For the gas in the two states we have

Initial state: $P_1 = 0.863$ atm; $V_1 = 11.21$; $n_1 = n$; $T_1 = T$; $P_1V_1 = nRT$

Final state: $P_2 = ?$; $V_2 = 15.01$; $n_2 = n$; $T_2 = T$; $P_2V_2 = nRT$

In this problem nRT is the same in the two states and so

$$P_1V_1 = P_2V_2$$

$0.863 \text{ atm} \times 11.21 = P_2 \times 55.01$

$$P_2 = \frac{0.863 \text{ atm} \times 11.21}{15.01}$$

$$= 0.644 \text{ atm.}$$

Example 12:

250 cm³ of a gas is collected over acetone at –10°C and 85 cm Hg. If the gas weighs 1.34 g and the vapour pressure of acetone at –10° is 39 mm Hg, what is the molecular mass of the gas?

Solution:

By Dalton's law

$$P = P_g + P_{acetone}$$

or $$P_g = P - \rho_{acetone}$$

$$= (850 - 39) \text{ mm Hg} = 811 \text{ mm Hg}$$

$$PV = nRT = \frac{g}{M} RT$$

or $$M = g.\frac{RT}{PV}$$

$$R = 0.0821 \frac{1 \text{atm}}{\text{mole} ^\circ K}$$

$$t = 273 - 10 = 263^\circ K$$

$$V = 250 \text{ cm}^3 = \frac{250}{1000} - 1 = 0.25 \ 1$$

$$P = 811 \text{ mm Hg} = \frac{811 \text{mm Hg}}{760 \text{ mm Hg}} \times 1 \text{ atm} = \frac{811}{760} \text{ atm}$$

$$g = 1.34 \text{ g}$$

Substituting,

$$M = \frac{1.34 \text{ g} \times 0.0832 \frac{1}{\text{mole} ^\circ K} \times 263^\circ K}{\frac{811}{760} \text{ atm} \times 0.251}$$

Example 13:

A lighter-than-air balloon is designed to rise to a height of 50 kilometres at which point it will be fully inflated. At that altitude the atmosphere pressure is 1.52 mm Hg and the temperature is –5°C. If the full volume of the balloon is one hundred thousand litres, how many kilograms of helium will be needed to inflate the balloon?

Solution

To calculate the number of kilograms of helium required we shall first calculate the number of moles and then convert these to kilograms.

The balloon has flexible walls, hence, the volume of the balloon varies with the applied pressure and at any height the pressure inside the balloon will be substantially equal to the atmospheric pressure. As the atmospheric pressure at a height of 50 km when the balloon is fully inflated is 1.52 mm Hg, the pressure inside the balloon will also be 1.52 mm Hg.

Hence, for the helium at a height of 50 km, we have,

$$P = 1.52\text{mm Hg}$$

$$V = \text{Volume of balloon when fully inflated}$$

$$= 100{,}0001$$

$$= 1.0 \times 10^{51}$$

$$n = ?$$

$$x = 0.821 \frac{1 \text{ atm}}{\text{mol}^\circ \text{K}}$$

$$T = 273–5 = 268°\text{K}$$

and $$n = \frac{PV}{RT}$$

Before substituting the given numerical values we have to see that the variables in the equation are expressed in the proper units. If we choose to use the value of R expressed in 1 – atm/mole°K, then the units of the terms in the equation are of necessity to those that appear in R and any quantity or quantities which do not have those units must be converted before substituting in the gas law.

As R is in –1 –atm/mole –°K, the pressure must be in atm, V in L and Tin kelvin.

Therefore, $P = 1.52 \text{ mm Hg} \times 1 - \text{atom}/760 \text{ mm}$

$$\text{Hg} = 2.0 \times 10^{-3} \text{ atm.}$$

$$V = 1.0 \times 10^{5} \text{ l and r} = 268°\text{K}$$

Substituting the numerical values

$$n = \frac{PV}{RT} = \frac{20 \times 10^{-3} \text{ atm} \times 1.0 \times 10^{5} \text{ 1}}{0.0821 - \dfrac{1-\text{atm}}{\text{mole}^\circ \text{K}} \times 269^\circ \text{K}}$$

$$= 9.090 \text{ moles.}$$

Since one mole of helium weighs 4 gm the mass of helium is

$$9.09 \text{ moles} \times \frac{4.00}{1 \text{mole}} = 36.36 \text{ gm}$$

$$= 0.03636 \text{ kilograms.}$$

Example 14:

The density of oxygen at 25°C is 1.43 g/1 at 1 atm pressure. At what pressure will oxygen have a density twice this value?

Solution:

To solve this problem we use the relation

$$P = \rho\frac{RT}{M}, \frac{P}{\rho} = \frac{RT}{M}$$

For the gas in the two states, we have

Initial state: $P_1 = 1$ atm; $1 = 1.43$ g/1; $T_1 = T$; $\frac{P_1}{\rho_1} = \frac{RT}{M}$

Final state: $P_2 = ?$; $\rho_2 = 2 \times 1.43$ g/1; $T_2 = T$; $\frac{P_2}{2} = \frac{RT}{M}$

Here $\frac{RT}{M}$ is the same in the two states and so

$$\frac{P_1}{\rho_1} = \frac{P_2}{\rho_2}$$

or $$P_2 = \frac{P_2}{\rho_1} \times P_1$$

Substituting the numerical values

$$P_2 = \frac{2 \times 1.43 \text{ g/1} \times 1 \text{ atm}}{1 \times 1.43 \text{ g/1}} = 2 \text{ atm.}$$

Example 15:

A sample of gas weighing 5.75g occupies a volume of 3.41 at 50° and 715 mm Hg pressure. What is the molecular mass of the gas?

Solution:

To solve this problem, we use the equation

$$P = \frac{gRT}{VM}$$

or $$M = \frac{gRT}{PV}$$

In this problem, we have

$g = 5.75gm$

$R = 0.0821 \frac{1 \text{ atm}}{\text{mole}^\circ K}$

$T = 273 + 50 = 323$ °K the pressure used must be in atm and V in 1

$P = 7.15mm\ Hg$

As R is $\frac{1-\text{atm}}{\text{mole}-^\circ K}$

$P = 715mm\ Hg$

$= 715mm\ Hg \times \frac{1 \text{ atm}}{760 \text{ mm Hg}}$

$= \frac{715}{760}$ atm

$V = 3.41$

Substituting the values,

$$M = \frac{5.5g \times 0.0821 \frac{1 \text{atm}}{\text{mole}^\circ K} \times 323^\circ K}{\frac{715}{760} \text{atm} \times 3.41}$$

$= 47.7$ g/mole.

Example 16:

One method for the commercial production of chlorine uses the electrolysis of molten sodium chloride. The chemical reaction that occurs is

$$2NaCl\ (l) \rightarrow 2Na(l) + Cl_2(g)$$

Solution:

How many litres of chlorine will be produced from one kg of molten sodium chloride if the gas is measured at 27°C and 1 atm pressure?

From the chemical equation, it is apparent that for each 2 mole ($2 \times 58.5 = 117.0$ g) of NaCl electrolysed, there will be formed 2 moles of Na(I) and one mole (35.5g) of chlorine. This information can be written as

$$2 \text{ moles NaCl} \rightarrow 1 \text{ mole } Cl_2$$

$$117 \text{ g NaCl} \rightarrow \text{mole } Cl_2$$

Therefore,

$$1000 \text{ g NaCl} \rightarrow \frac{1\,\text{mole}}{117\,\text{g}} \times 1000 \text{ g } Cl_2$$

$$n = 8.55 \text{ moles } Cl_2$$

Assuming that chlorine obeys ideal gas law,

$PV = nRT$, we have

$$V = \frac{nRT}{P}$$

$$R = 0.0821 \frac{1 \text{ atm}}{\text{mole } ^\circ K}$$

$$P = 1 \text{ atm}$$

$$T = 273 + 27 = 300^\circ K.$$

Hence, $$V = \frac{8.55 \text{ mole} \times 0.082 \frac{1\,\text{atm}}{\text{Mole}^\circ K} \times 3000^\circ K}{1 \text{ atm}}$$

$$= 2.106 \times 10^2 1.$$

Example 17:

The density of hydrogen at 0°C and 760 mm Hg pressure is 0.00009 g/cm³. What is the root mean square speed of hydrogen molecules?

Solution:

$$PV = \frac{1}{3} m N \overline{u}^2$$

Therefore, $$\overline{u}^2 = 3\frac{PV}{mN}$$

As m is the mass of a molecule and N is the number of molecules, therefore, mN, the total mass; so = Mass/Volume = mN/V

and $$\overline{U}^2 = \frac{3P}{\rho}$$

$$\text{or} \quad U_{rms} = \left(\overline{U^2}\right)^{1/2} = \left(\frac{3P}{\rho}\right)^{1/2}$$

But $\quad P = 76 \times 13.59 \times 981 \text{ dynes/cm}^2$

$$= \left(\frac{3 \times 76 \times 13.50 \times 981}{0.0009 \text{ g/cm}^3} \text{ dynes/cm}^2\right)^{1/2}$$

$$= 183.8 \times 10^8 \text{ cm/sec} = 1.838 \text{ km/sec.}$$

Example 18:

A tube 1 metre long and 10.0 cm² cross section is filled with nitrogen at a pressure of .54 atm at 25°. The ends are then sealed with hydrogen permeable material. One end is put in contact with a source of hydrogen at STP while the other end is in contact with vacuum. A steady state is obtained. Estimate the flow rate of hydrogen.

Solution:

$$\frac{1}{A} m_k = -D_k \frac{C_k(2) - C_k(1)}{D}$$

$$\frac{1}{10\text{cm}^2} m_k = \frac{.67\text{cm}^2/\text{sec} \times .54 \text{ atm}}{1 \text{ metre}}$$

$$m_k = 0.0217 \text{ y/hour.}$$

Example 19:

Calculate the kinetic energy of an Avogadro's number of molecules of gas molecules (that is one mole of gas) at 0°C.

Solution:

Kinetic energy of Avogadro's number of molecules $= N_o \overline{KK}$

$$= N_o \cdot \frac{3}{2} RT \left(R = kN_o\right)$$

The value of R is 1.987 cal/mole °K, and T 273°K, thus, kinetic energy of Avogadro's number of molecules

$$= \frac{3}{2} 1.987 \frac{\text{cal}}{\text{mole.K}} \times 273 \text{ K}° = 814 \text{ cal/mole.}$$

Note that in this calculation, the nature of gas molecules is unimportant. All ideal gases at 0°C will have this amount of kinetic energy.

The average kinetic energy of gas molecule $\left(\overline{KE}\right)$ is generally expressed in ergs and not in calories and at 0°C, it has the value.

$$\overline{KE} = \frac{3}{2} KT$$

$$= \frac{3}{2} \times 1.381 \times 10^{-16} \text{ erg/molecule K} \times 273°K$$

$$= 5.66 \times 10^{-14} \text{ erg/molecule.}$$

Example 20:

Calculate the root mean square speed of oxygen molecules in the lungs at normal body temperature 37°C.

$$u_{rms} = \left(\frac{3RT}{M}\right)^{1/2}$$

Solution:

In order to give U_{rms} the units of cm/sec, the value of R must be expressed in mechanical energy units, *i.e.*, in ergs (1 erg equal 1 g.cm²/ sec²).

$$R = 8.31 \times 10^7 \text{ ergs/mole}$$

$$°K = 8.31 \times 10^7 \text{ g.cm}^2 \text{ sec}^{-2} \text{ mole}-^1 °K^{-1}$$

$$T = 273 + 37 = 310°K$$

$$M = 32 \text{ g/mole}$$

Substituting these values in the equation

$$u_{rms} = \left(\frac{3 \times 8.31 \times 10^7 \text{ g.cm}^2 \text{ sec}^{-2} \text{ mole}^{-1} °K \times 310° K}{32.0 \text{ g.mole}^{-1}}\right)^{1/2}$$

$$= 4.92 \times 10^4 \text{ cm sec}^{-1}.$$

Example 21:

Calculate the pressure exerted by 5.0 moles of carboy dioxide in a 1-litre flask at 47°C, using: (a) the Van der Waals equations, and (b) the ideal gas kw. Compare these results with the experimentally observed pressure of 82 atm.

Solution:

(a) The Van der Waals equation is,

$$\left(P + \frac{n^2a}{V^2}\right)(V - nb) = nRT$$

P = ?

n = 5.0 moles

V = l litre

r = 47 + 273 = 320°K

R = 0.0821 l.atm $mole^{-1}$ $°K^{-1}$

From the table

a = 3.592 $litre^2$ atm. $mole^{-2}$;

b = 0.0427 litre $mole^{-2}$

Substituting these values

$$\left(P + \frac{25 \times 3.592}{2}\right)(1.0 - 5.0 \times 0.0427) = 5.0 \times 0.0821 \times 320$$

$$(P + 25 \times 3.592)(0.7865) = 5.0 \times 0.0821 \times 320$$

$$P + \frac{5.0 \times 0.0821 \times 320}{0.7865} - 25 \times 3.92 = 77.2 \text{ atm.}$$

(b) For an ideal gas, PV = nRT

$$P \times 1.0 = 5.0 \times 0.0821 \times 320$$

$$P = 5.0 \times 0.0821 \times 320 = 131.3 \text{ atm.}$$

The pressure calculated from the Van der Waals equation is closer to the experimental value (85 atm). This is to be expected because the total pressure of the gas is high enough to cause significant deviations from the ideal behaviour.

Example 22:

Two parallel plates are each of 100.0 cm^2 area and are spaced 1.0 cm apart. One plate is kept at 10.0°C by some means, the other plate at 0°C by being in thermal contact with a mixture of ice and water. The region between the plates is filled with air at atmospheric pressure (K is 0.586 × 10^{-4} calcm sec. °C). The latent heat of fusion of ice is 79.7 cal/g. Calculate the rate of melting of ice in the freezing mixture.

$$\frac{dQ}{dT} = AK\left(\frac{dT}{dz}\right) = 100 \text{ cm}^2 \times 0.586 \times 10^{-4} \text{ cal/cm sec°C}$$

$$\times 10°C \times 60 \text{ sec}$$

$$= \frac{3516 \times 10^{-2}}{797} \text{ g/min} = 0.0441 \text{ g/min.}$$

Example 23:

Argon gas is used in the welding of stainless steel sheets. Calculate its $\left(\overline{C^2}\right)^{1/2}$, $\overline{C}$ *and* C_p *at 25°C.*

Solution:

$$\left(\overline{C^2}\right)^{1/2} = \left(\frac{3\,RT}{m}\right)^{1/2}$$

$$= \left(\frac{3 \times 8.314 \times 10^7 \times 298}{40}\right)^{1/2}$$

$$= 4.32 \times 10^4 \text{ cm sec}^{-1}$$

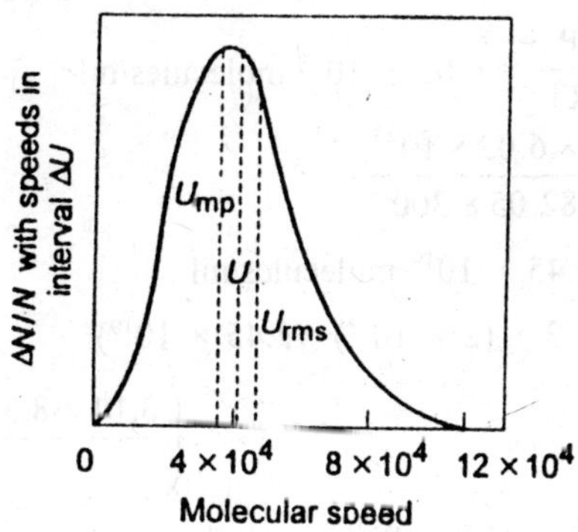

Fig. 3.52

$$\overline{C} = \left(\frac{8\,KT}{\pi\,m}\right)^{1/2} = \left(\frac{8\,RT}{\pi\,m}\right)^{1/2} \left(\frac{8 \times 8.314 \times 10^7 \times 298}{3.14 \times 40}\right)^{1/2}$$

$$= 3.98 \times 10^4 \text{ cm sec}^{-1}$$

$$C_p = \left(\frac{2\,RT}{M}\right)^{1/2} = \left(\frac{2 \times 8.314 \times 10^7 \times 298}{40}\right)^{1/2}$$

$$= 3.52 \times 10^4 \text{ cm sec}^{-1}$$

Example 24:

Calculate the most probable velocity for electron at 25°C.

Solution:

$$C_p = \left(\frac{2 \times 1.38 \times 10^{-16} \times 298}{9.108 \times 10^{-28}}\right)^{1/2}$$

$$= \left(\frac{2 \times 1.38 \times 298}{9.108}\right)^{1/2} \times 10^{6} \text{ cm sec}^{-1}$$

$$= 9.504 \times 10^{6} \text{ cm sec}^{-1}.$$

Example 25:

Calculate the total number of bimolecular collisions in 1 sec in 1 ml of pure He and pure N_2 at 300°K at 1 atm. The collision diameters of He and N_2 are 2×10^{-8} cm and 4×10^{-8} cm respectively.

Solution:

$$n = \frac{P}{RT} \times 6.02 \times 10^{23} \text{ molecules/ml}$$

$$= \frac{1 \times 6.02 \times 10^{23}}{82.05 \times 300}$$

$$= 2.45 \times 10^{19} \text{ molecules/ml.}$$

$$Z_{12} (\text{He}) = 2 \times (2 \times 10^{-8})^2 (2.45 \times 10^{19})^2 \left(\frac{3.14 \times 8.314 \times 10^{7} \times 300}{4}\right)^{1/2}$$

$$= 7 \times 10^{28} \text{ collisions sec}^{-1} \text{ mr}^{-1}$$

$$Z_{12} (N_2) = 2(4 \times 10^{-8})^2 (2.45 \times 10^{-9})^2 \left(\frac{3.14 \times 8.314 \times 10^{7} \times 300}{28}\right)^{1/2}$$

$$= 10 \times 10^{28} \text{ collisions sec}^{-1} \text{ mr}^{-1}.$$

Example 26:

Calculate the mean free path of hydrogen and oxygen at 1 atm pressure and 25°C.

Solution:

$$\frac{m}{V} = (4.89 \times 10^{-5})\,(6.02 \times 10^{23}) = 2.46^2 \times 10^{-19} \text{ mole/c.c.}$$

$$\lambda_{H_2} = \frac{1}{(2)^{1/2}} \times 3.14 \times (2.74 \times 10^{-8})^2\,(2.462 \times 10^{-19})$$

$$= 1.217 \times 10^5 \text{ cm.}$$

$$\lambda_{O_2} = \frac{1}{(2)^{1/2}} \times 3.14 \times (3.61 \times 10^{-8})^2\,(2.462 \times 10^{-19})$$

$$= 7.015 \times 10^{-6} \text{ cm.}$$